TRANSPORTATION RESEARCH
RECORD

No. 1474

Soils, Geology, and Foundations

Mechanically Stabilized Backfill and Properties of Geosynthetics and Geocomposites

A peer-reviewed publication of the Transportation Research Board

TRANSPORTATION RESEARCH BOARD
NATIONAL RESEARCH COUNCIL

NATIONAL ACADEMY PRESS
WASHINGTON, D.C. 1995

Transportation Research Record 1474
ISSN 0361-1981
ISBN 0-309-06110-5
Price: $27.00

Subscriber Category
IIIA soils, geology, and foundations

Printed in the United States of America

Transportation Research Record 1474

Contents

Foreword

The 13 papers included in this volume are arranged into two general groups. The first group consists of papers that are related to mechanically stabilized backfill (MSB) materials. These papers focus on advanced technologies related to MSB and on ownership roles in providing support during the design and construction phases. These papers also discuss performance of MSB applications using case histories.

The second group of papers is on properties of geosynthetics and geocomposites. These papers include information on the long-term durability of geosynthetics used as soil reinforcements; the frictional mechanism of geogrid-soil systems on the basis of results from tests on different types of geogrid-soil combinations; and the properties of granular and clayey soils that are reinforced using multioriented geosynthetic elements, staple fiber, continuous filament, and synthetic and steel fibers.

PART 1

Mechanically Stabilized Backfill

Texsol: Material Properties and Engineering Performance

PHILLIP LIAUSU AND ILAN JURAN

Texsol is a composite material made of sand and continuous polyester fibers mixed together in situ to form a homogeneous construction material. The fiber content varies between 0.1 and 0.2 percent of the weight of sand. The fibers provide for the high cohesion of Texsol and its ability to sustain large strains without degradation of its mechanical properties. The sand is well-graded medium course material and provides for the internal friction resistance of Texsol and its self-draining characteristics. Substantial testing programs have been conducted by state agencies, universities, and research institutions in France and subsequently in Japan to assess the engineering performance of this composite material and develop relevant design methods for its various fields of application.

Texsol is a composite material made of sand and continuous polyester fibers mixed together in situ to form a homogeneous construction material. The fiber content varies between 0.1 and 0.2 percent of the weight of sand. The fibers provide for the high cohesion of Texsol and its ability to sustain large strains without degradation of its mechanical properties. The sand is well-graded medium course material and provides for the internal friction resistance of Texsol and its self-draining characteristics.

Substantial testing programs have been conducted by state agencies, universities, and research institutions in France and subsequently Japan to assess the engineering performance of this composite material and develop relevant design methods for its various fields of application. The research and development programs, as well as field observations on more than 100 Texsol structures constructed since 1984, demonstrated that the engineering properties of Texsol include high shear resistance with anisotropic mechanically built-in internal cohesion and internal friction angle that are dependent on the fiber content (1,2), self-draining properties of the sand used, low creep potential under normal operating conditions, durability and sustainable resistance to chemical and biological attacks, high ductibility and large energy absorption capacity with high resistance to impact, explosions, and seismic effects (3,4); deformability and large tolerance to differential settlements; high resistance to runoff surface erosion (5), and high thermal resistance under fire-generated heat up to 600°F (6). In addition, Texsol provides a suitable support for plant roots to penetrate and seeds to germinate. Mixed in organic soil, fertilizer, and seeds, the Texsol green method enables the hydroseeding of steep natural slopes, excavated slopes, embankments, retaining walls, soundproof walls, and so forth, where conventional hydroseeding techniques are impractical.

Because of its remarkable features, Texsol has been increasingly used in a variety of engineering applications (Figure 1), including earth retaining walls, particularly on soft compressible soils, with facing slope angles of 65 to 75 degrees; stabilization of earth slopes in cuts and embankments; steepening of existing slopes for widening of motorways; surface protection of man-made and natural slopes against rock falls and surface erosion due to climate conditions (e.g., freezing temperature); and explosion-resistant facilities in civil and military installations for storage of explosives and liquefied gas, offering a remarkable market potential for civil engineering construction in earthquake zones.

This paper presents the main results of the research conducted to assess the material properties and engineering performance of Texsol's structural applications.

MATERIAL PROPERTIES

Sheer Strength Characteristics

The mechanical properties of Texsol depend on the characteristics of the granular material used, thread type, fiber content, production equipment, and compaction parameters (density and water content).

Figure 2 shows the results of triaxial compression tests performed on samples of Texsol and unreinforced sand under different confining pressures and the related characteristic failure curves of these materials. Test results illustrate that the shear modulus of Texsol and its hydraulic conductivity are similar to that of the natural sand. The main mechanical properties of Texsol are

- Unconfined compressive strength: 500 kPa/0.1 percent of fill content ratio by weight;
- Apparent cohesion of 100 kPa/0.1 percent of fill content ratio by weight;
- Internal friction angle that is equal or greater than that of the natural sand, with
  $$\varnothing \text{ Texsol} = \varnothing \text{ soil} + \Delta\varnothing,$$
  ($\Delta\varnothing$ varies from 0 to 10);
- Yield strain that is greater than that of the natural sand, indicating the ductile behavior of Texsol with
  $$E^{\text{Texsol}} = E^{\text{soil}} + \Delta E^r$$
  (ΔE varies from 0 to 10 percent);

Because of the production process of Texsol, the shear strength characteristics are anisotropic, that is, function of the inclination angle $\propto$ of the shear failure surface with respect to the depositional plane of the material. Figure 3 shows the results of direct shear tests of Texsol specimens prepared with a reference 0/5 mm sand, polyester fiber with a linear density of 167 define, and fiber content of 0.2 percent by weight, prepared at the normal proctor density. The results

P. Liausu, Menard Sol Traitement, P.O. Box 530, 91946, Les Uli Cedex France. I. Juran, Department of Civil and Environmental Engineering, Polytechnic University, Brooklyn, N.Y. 11201.

Retaining wall in Asterix theme park Plailly (before seeding)

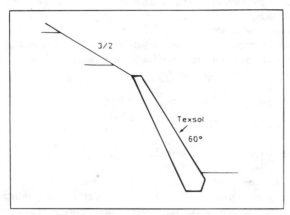

Widening of Highway A12 - Bois d'Arcy

FIGURE 1   Examples of Texsol engineering applications.

illustrate the effect of the inclination angle $\propto$ on the apparent cohesion C_t and the internal friction angle $\varnothing^c$ of the reference Texsol material. With the present state of knowledge, the anisotropy of $\varnothing t$ is not taken into account and $\varnothing^t$ is assumed to be constant and equal to the friction angle of the natural granular material which results in a conservative design. The anisotropy of the apparent cohesion of Texsol follows the empirical equation derived from the analysis of the test results obtained for the reference Texsol material

$$C_t = 0.03 \propto^2 + 1{,}27 \propto + 16.5 \text{ (in kPa)}$$

Creep Behavior and Durability Consideration

Creep behavior of construction materials must be considered in civil engineering the result of permanent load and long life duration of constructions. In the case of geotextile reinforcement, creep studies have been made in order to select the proper reinforcing material and to evaluate the long term deformations to be expected.

A first conclusion of that research is that creep effects depend on polymer type. Polymers are characterized by their glass transition temperature Tg. Tg of polyester is around 79°C and Tg of polyolefins is below 0°C. As soil-structure temperatures are usually in

the 0° to 30°C range, the basic difference between these materials will affect their engineering behavior. Below the Tg temperature, the polymer is a solid and will creep only under high working loads; above that temperature, the polymer will creep even under low working loads. This fundamental difference between the polymers has been the prime reason for the selection of polyester thread for Texsol structures such as retaining walls that have to sustain permanent loading. Typical characteristics of the polyester fibers currently used in Texsol structures are indicated in Table 1.

Creep effects result in both a reduction of failure strength (resulting from long-term loading) and long-term strain. It has therefore been necessary to demonstrate that the polyester fiber–reinforced Texsol material is not affected by creep under the working loads generally used in civil engineering structures. To address these issues, two series of creep tests have been conducted.

The first series of tests consisted of four long-term simple compression tests at room temperature. Two Texsol samples with a 0.12 percent proportion of polyester thread were submitted to 60 percent of their failure strength (as determined from another series of tests on reference samples) during 2.5 years, and two additional samples were loaded at 45 percent during 3 years. The rate of strain under the 60 percent load, after the initial settlement, has been linear with respect to log (t) with a slope smaller than 10-2 per cycle (i.e., less than 2 percent axial strain between 1 and 100 years). The samples loaded at 45 percent gave a strain rate of 5.10-3 per cycle of log (t). One sample loaded at 60 percent has been tested under compression after 850 days; the measured strength was equivalent to the short-term strength of the reference samples. These tests yield two important indications:

- Time-dependent deformations of the composite material made of polyester thread and granular material, for a given working load (as determined for the composite material itself), are significantly smaller than creep deformations measured on the thread alone for the same working load (as determined for the thread).
- Measured rate of strain, whether due to polymer creep, remains very low and does not generally need to be considered in geotechnical design of conventional retaining structures for fills and cuts.

The second group of creep loading tests has been done, at an elevated temperature (50°C or 60°C) to accelerate the creep under laboratory-controlled conditions. After triaxial loading at an elevated temperature, the samples were tested up to failure at ordinary temperature to measure their strength after preloading. This testing program included 50 samples. Preloading has been at two-thirds of the failure strength for most samples.

These tests have resulted in three main conclusions.

- The rate of strain under constant load has been found to be around 5.10-3 per cycle of log (t) at both 50°C and 60°C, which is close to the strain rate obtained under room temperature. Therefore, it is anticipated that this time-dependent deformation is not due to creep, because it is temperature independent. This deformation may be the result of the sand consolidation.
- As mentioned previously, the measured rate of time-dependent deformation can be ignored for most applications of the material (2 percent between 1 year and 100 years).
- Material strength is not decreased by the preloading: measured strength values after the loading period are equal or higher than the reference values determined on nonpreloaded samples. These

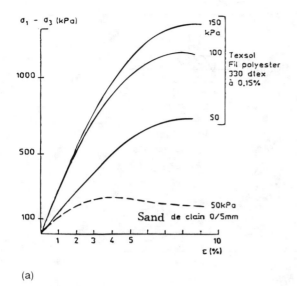

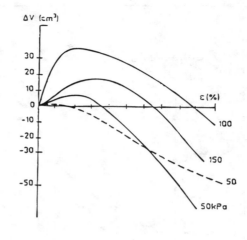

(a)

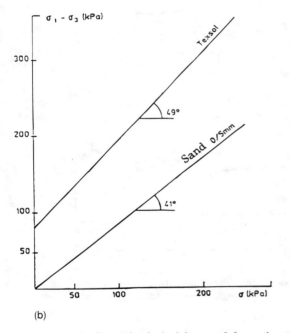

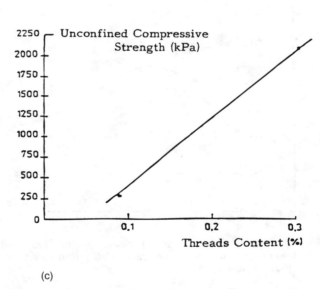

(b) (c)

FIGURE 2 **(a) Example of triaxial stress-deformation curves, (b) failure envelope for Texsol and natural sand, (c) Texsol T1: compressive strength versus threads proportion.**

results further support the assumption of sand consolidation–induced deformation.

Degradations have been observed, which can be explained by mechanical stresses (compression, shear, abrasion), by ultraviolet light action, in case of long-term exposure, or by the influence of specific environments, such as cement during setting. However, the experience of more than 20 years with polyester geotextile structures illustrates that for fibers embedded into the soil mass, in most cases no chemical changes have been detected internally or on the surface of fibers. Furthermore, the statistical study of pH values of granular materials that can be used for Texsol shows that for the range of temperatures and pH values that are likely to be encountered in the natural environment, risk of hydrolysis degradation is not to be considered in design practice. However, the use of granular industrial wastes as a constituent of Texsol or applications in the presence of very specific industrial environments would require an appropriate investigation, which would also be routinely required if concrete, steel, or other materials are used. For extreme situations, different types of polymers could be used.

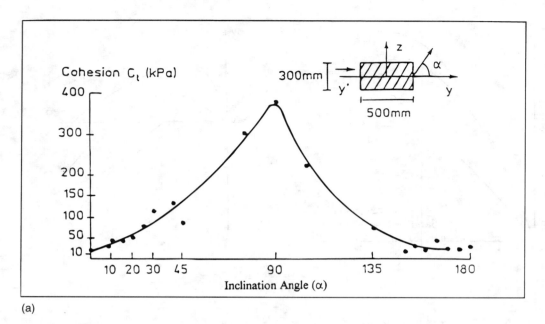

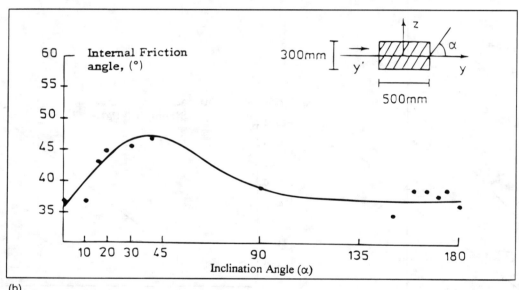

FIGURE 3   (a) Anisotropy of Texsol cohesion: experimental values from direct shear test, (b) internal friction angle of Texsol: experimental values from direct shear test (α—angle between layering plane and shearing plane).

TABLE 1   Typical Characteristics of Polyester Fiber used in Texol

Nature	Type of fill	Title (dtex)	Number of thread	Tenacity (cN/tex)	Extent. at failure (%)	Initial Modulus (cN/tex)
Polyester PES	Thread (integrated extrusion)	50	16	40	25	970
		167	30	36	26	770
		330	60	37	27	950
		280	60	58	19	790
		280	48	61	14	800

In situ observations and related laboratory analyses did not indicate any biological effect on the stability of polyester fibers. Furthermore, standard laboratory tests, using soils with a known bacteria content, indicated no reduction in the strength of polyester threads used for Texsol. It can therefore be concluded that Texsol has high durability and sustainable resistance to both chemical and biological attacks.

Dynamic Response Properties

Present knowledge on the seismic behavior of Texsol results from cyclic laboratory tests performed in France and both model studies and a full-scale experiment on a Texsol structure conducted in Japan.

The laboratory tests performed by Luong included

- Conventional triaxial tests with constant confinement, monotonic loading, repeated loading, long-duration cyclic loading, and large strain loading;
- Constant mean stress triaxial tests with monotonic loading and cyclic loading;
- Tests with monotonic or cyclic lateral loading;
- Liquefaction tests; and
- Longitudinal resonance tests and behavior under vibrations.

Impact behavior and wave propagation have also been considered.

The triaxial tests have shown the high ductility of Texsol and its high energy-absorption capacity, resulting from the high dilatancy occurring between the critical state (zero volumetric strain) and the peak strength. Energy absorption results from friction between particles of the material when deformation develops; the threads in the Texsol material allow large strains to exist in the granular material before failure; consequently, high energy absorption is possible while keeping a sufficient safety margin with respect to failure.

Cyclic compression and extension triaxial tests on reference Texsol material was performed (7) with a Fontainebleau sand, polyester 50/16 of 50 dtex with a fiber content of 0.2 percent by weight. For the high cyclic loading amplitude that exceeded the critical state (zero volumetric strain line), the test results illustrated in Figure 4 demonstrate a progressive densification of the Texsol material with the increasing number of cycles.

The liquefaction potential of Texsol has been investigated through cyclic triaxial deviatoric load testing, with cyclic loading amplitude exceeding the critical state line both in compression and extension. As illustrated in Figure 5, after a number of cycles, Texsol liquefaction tests show a stabilization of the stress-strain cycles indicating a high energy absorption resulting in high liquefaction resistance.

ENGINEERING PERFORMANCE OF TEXSOL WALLS

Static Loading

Several full-scale experiments have been conducted by the Regional Laboratory of Rouen in France to assess the engineering performance of Texsol walls. Figure 6a shows the cross section and site characteristics of the experimental wall, 3 m high with a facing inclination of 68 degrees, retaining an unreinforced Fontainebleau sand fill that was loaded up to failure. Figure 6b shows the facing displacements during the loading, illustrating a progressive rota-

tional failure mechanism. The displacement records indicate that the surcharge loadings should exceed 75 percent of the failure loading to generate significant lateral displacements.

Seismic Loading

Tests performed in Japan (4) in cooperation with the National Research Institute of Agricultural Engineering (Ibaraki, Japan) and Kumagai Gumi Co., Ltd., included (a) a series of shaking table tests on models of earth dam with reinforced facing and (b) a 10-m-high test wall retaining an earth fill instrumented to evaluate its response to natural earthquakes.

Earth embankment models, 0.4 or 0.8 m high, were made of loose sand (with no impervious layer) with a downstream horizontal drain and tested with an upstream water level equal to three-fourths of the embankment height. Model facings were made of loose sand or reinforced with a compacted sand layer or a Texsol layer 10 to 15 cm thick. Models 0.4 m high were submitted to an input sine wave with a frequency of 10 Hz and with acceleration levels of 100, 200, 400, and 600 gal, applied during 10 sec. Models 0.8 m high were submitted to a 3-Hz vibration with acceleration levels of 150, 250, and 450 gal.

The parameters measured were acceleration, pore pressure, and settlement. Settlement of the crest and continuity of strains were considered indications of the effectiveness of the reinforcement method because they are critical to the risk of overflow. The models demonstrated that the use of Texsol significantly reduced settlements and created no cracks.

Figure 7 compares the settlements of the crest observed on the 0.4-m-high model under three conditions: unreinforced, reinforced with a dense sand layer, and reinforced with a Texsol layer. Four sec after loading, a settlement of approximately 25 mm occurred in the unreinforced model, but almost none occurred in the model reinforced with Texsol fibers. Seven sec after loading, the settlement of the model reinforced with continuous fibers (compared with the unreinforced one) was reduced to approximately one-third.

Figure 8 compares the settlement of the crest observed on the 0.8-m-high model at a 450-gal input. The crest settlement in the unreinforced embankment reached approximately 4 cm, and resulting cracks developed over the entire model embankment. In the model reinforced with the continuous fibers, almost no settlement occurred. The results of these large-scale shaking table tests demonstrated the effectiveness of the continuous fiber reinforcement.

The 0.8-m-high model with Texsol had a maximum settlement of 6 mm, without cracks, whereas the unreinforced model showed a 41-mm settlement, with cracks propagated over the entire model, resulting in its collapse.

The 10-m-high wall illustrated in Figure 9 was monitored under natural conditions for a long-term performance evaluation. The retaining wall was completed in December 1988; since that date, it has undergone heavy rains, typhoons, and earthquakes up to a magnitude of 5.7 on the Richter scale (February 19, 1989). The wall showed no damage and stability was maintained.

The outer slope of the wall is 1:0.5 (63 degrees horizontally); the width at the base is 2.5 m and the width at the top is 1 m. The retained fill material has a density of 15.9 kN/m³, a water content of 50 percent, a cohesion of 6 kPa, and an angle of internal friction of 18 degrees. During the February 19, 1989, earthquake, the measured acceleration at the ground surface perpendicular to the axis of the wall was 95 gal; the power spectrum showed accelerations from

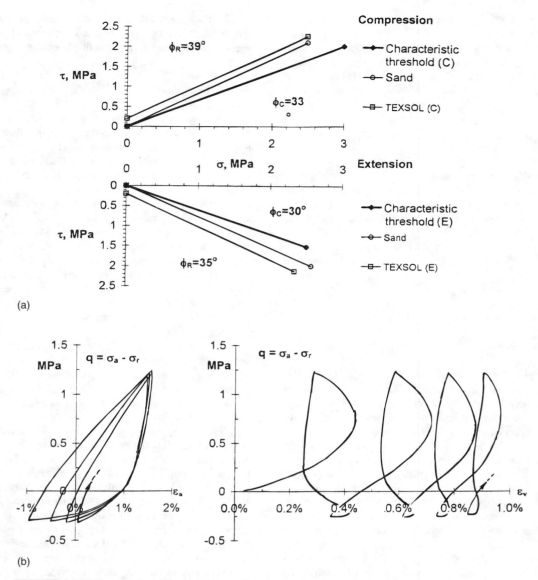

FIGURE 4   (a) Texsol triaxial compression and extension tests; (b) densification behavior under high-amplitude repeated loading.

2 to 8 Hz. Vibration measurements on the wall allowed evaluation of its dynamic behavior (natural period around 0.4 sec).

Measurements of earth pressures on the wall between the fill and the Texsol material at different heights showed large variations of earth pressure during the earthquake because of the deformability and inertia of the wall. Figure 9 shows the cross section of the wall, its instrumentation, and the distribution of the maximum increase of the earth pressure during the earthquake. The observed distribution of the earth pressure increases from the static level to the maximum value as the earthquake is compared with the calculated values obtained based on the Mononobe-Okabe formula commonly used for earthquake-resistant design. This comparison indicates that the experimental distribution of the earth pressure generated by the seismic effect is not a triangular distribution, and it differs considerably from the distribution computed by the Mononobe-Okabe formula.

The major observation made during this natural earthquake was that, although the static safety of the wall was already at a critical state, no damage was found.

Resistance to Surface Erosion of Retaining Structures

Texsol constructions can be subjected to a large spectrum of erosion conditions according to type of structure, normal or exceptional operating conditions, local climate, and types of hydraulic attacks for which it is designed.

As an example, the use of Texsol in a bank protection system, possibly with other techniques or materials, does not require the study of the same mechanisms as does use in retaining structures. For walls, the surface erosion evaluation attempts to establish whether a progressive loss of granular material could occur at the surface of the Texsol material from rain and wind. Such a loss could result in a slow reduction of wall thickness.

Observations of existing walls before grassing or grassing by simple hydroseeding (without application of the Texsol green method) indicate that the effect of weathering on the surface of Texsol retaining structures does not result in continuous erosion of the wall beyond the construction phase and periods of rain occurring

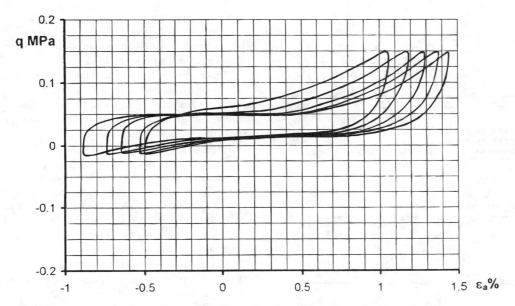

FIGURE 5   Liquefaction of saturated Texsol under controlled axial strain.

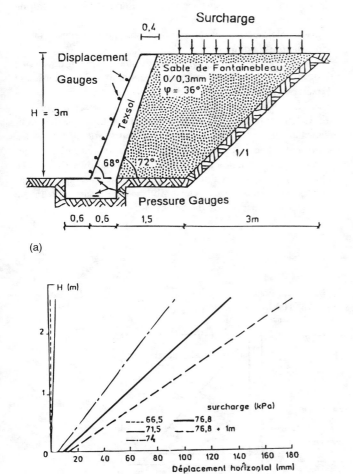

(a)

(b)

FIGURE 6   (a) Cross section of experimental wall, (b) successive deformation of wall No. 2 facing (8).

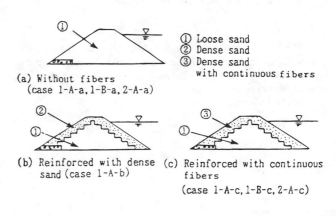

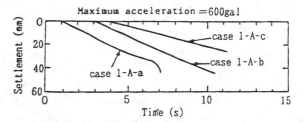

FIGURE 7   Small-scale shaking table test models: settlement of model crest (4).

shortly afterward. For example, such observations have been made on the many walls built along the A7 motorway in France. The longitudinal concrete surface collectors and drains placed in the cut areas where Texsol walls are built show no accumulation of sand material, which would have been carried from the toe of the walls toward the storm sewer.

In the course of a testing program on Texsol conducted by the Japanese Ministry of Construction, measurements have been made

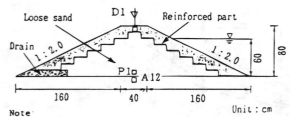

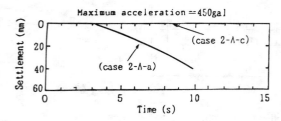

FIGURE 8   Large-scale shaking table test model (Model 2): settlement at D1 (Model 2, Sand A).

on a 5-m-high wall with a 63-degree slope, without vegetation. The wall was submitted to an intense artificial rain (30 mm/hr) and the amounts of eroded sand were measured. Translated into an average eroded thickness on the area of the face, the measured erosion showed the following values:

Total Rain (mm)	Eroded Thickness (mm)
100	5
200	9
300	12
400	14
500	15.2
600	16

It appears that the first rain periods have a washing effect on the surface, but that fairly quickly the erosion process slows down and, for all practical purposes, stops; extrapolating the experimental curve leads to a limiting value of 18 mm.

These measurements correspond well with observations made on actual projects, where the washed-out thickness occurring at an early age has been estimated, in temperate climate conditions, at an order of magnitude of 1 cm. However, incidental degradations have been observed as a result of a locally heavy running water flow: for example, the outflow at the top of a wall of a storm sewer resulted in local erosion important enough to require repair. Such surface water flows must be avoided. In particular, walls located below a large catchment area must be protected by an interceptor trench on top of the wall ensuring that an unknown quantity of water will not flow over the structure. Considering the relatively low rigidity of Texsol constructions, the trench is lined preferably with a material that will not crack, such as a geomembrane.

The absence of erosion under the action of rain is related to the intricate texture of the thread network contained in the Texsol material; in addition to this network knitted into the mass of the composite, the production process of Texsol often results in a superficial layer of threads oriented toward the slope and having weak connections with the material itself. All these threads are responsible for the erosion resistance that is observed, but the resulting appearance is often unsatisfactory when there is no vegetation or when simple grassing by seeding does not find sufficiently favorable growing conditions.

For this reason the Texsol green method has been developed (Figure 10) to reestablish appropriate conditions for a dense and durable vegetative cover, provided a proper water supply is available. The Texsol green method is used for the hydroseeding of Texsol walls or natural slopes where conventional hydroseeding techniques are impractical (e.g., excavated slopes, soundproof walls, steep embankments, etc.). It consists of Texsol mixed with fertilizer seeds and a coagulation agent, which is sprayed over the surface area of the structure. Artificial mesh is sometimes required to initially hold the Texsol green. Generally, the natural growth will take place gradually depending on the environment.

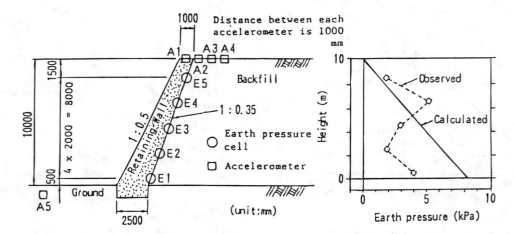

FIGURE 9   Model retaining wall and observation points: maximum increments of earth pressure due to earthquake.

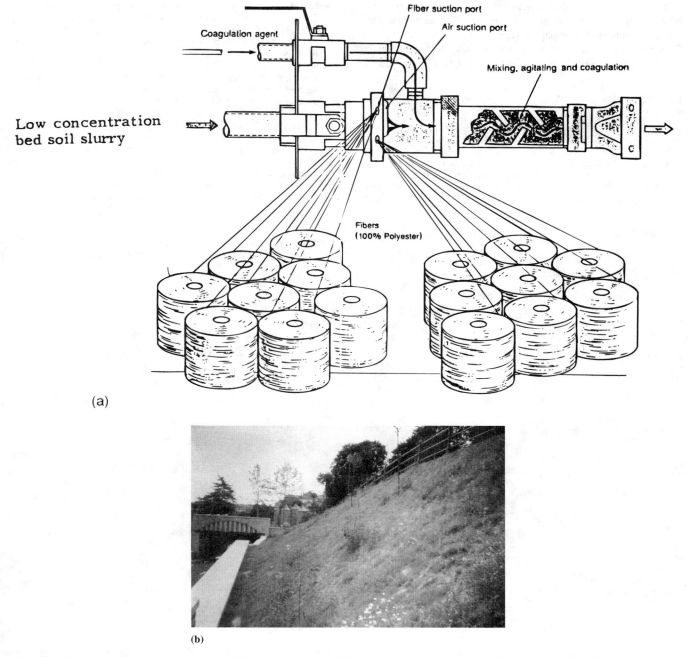

Low concentration
bed soil slurry

(a)

(b)

FIGURE 10   (a) Schematic of green method of Texol production; (b) typical construction site using Texsol green method.

In addition to its landscaping purpose, application of the Texsol green method (through the additional layer it gives, its specific layer of thread, and the root network of the grass cover it generates) introduces an additional resistance to surface erosion. Using this technique is therefore advisable, not only on Texsol walls but also on natural slopes with soils or rocks prone to weathering and corrosion.

CONCLUSION

The testing programs conducted in France and Japan to assess the engineering performance of Texsol structures have demonstrated that the reinforcement of sand by continuous polyester fiber pro-

vides the composite material with apparent cohesion, ability to sustain large strains, and high energy-absorption capacity that make Texsol structures a cost-effective solution for highway retaining systems under difficult site conditions, such as compressible soft soils and earthquake zones.

As for durability, the experience gained from geotextiles made of polyester is applicable to the Texsol material. The creep studies on the Texsol material have demonstrated that

• Under normal operating conditions, the Texsol material using polyester thread does not creep; and
• There is no decrease of strength with loading time.

It can also be stated that current erosion protection experience with Texsol illustrates that the use of the Texsol green method permits environmentally compatible vegetative structural surfaces for Texsol walls and man-made and natural slopes, while significantly increasing their resistance to weathering and surface erosion.

REFERENCES

1. Leflaive, E., and P. Liausu. Le Renforcement des Sols par Fils Continus. *Proc., 3rd International Congress on Geotextiles*, Vienna, Austria, April 1986, pp. 523–528.
2. Gigan, J. P., M. Khay, J. Marchal, M. Ledelliou. Proprietes Mechaniques du Texsol: Application aux Ouvrages de Soutenement. *Proc., 12th International Conference on Soil Mechanics*, Rio de Janiero, Brazil, Aug. 1989, pp. 1251–1252.
3. Luong, M. P., E. Leflaive, and M. Khay. Proprietes Parasismiques du Texsol. *Proc., 1st National Colloquiam on Antiseismic Engineering*, Vol. 1, Saint-Remy-Les-Chevreuses, France. Jan. 1986, pp. 3-49–3-58.
4. Fukuoka, M., et al. Stability of Retaining Wall Reinforced by Continuous Fibers During Earthquakes. *Proc., 4th International Congress on Geotextiles*, the Hague, the Netherlands, 1990, pp. 546–558.
5. Kutura, K., et al. Behavior of Prototype Steep Slope Embankment Having Soil Walls Reinforced by Continuous Threads. *Proc., 4th Geotextile Symposium*, Tokyo, Japan, Dec. 1989, pp. 335–342.
6. *Stabilité au Feu du Materiau Texsol*. Totalgaz internal report, 1990.
7. Luong, M. P. *Recherches sur le Comportement du Texsol sous Sollicitations Dynamiques: Comportement Transitoire et Dynamique du Texsol*. Ecole Polytechnique, Laboratoire de Mechanique des Solides, France, 1986.
8. *Texsol: Retaining Structures*. LCPC technical guide, 1990.

Publication of this paper sponsored by Committee on Geosynthetics.

Independent Facing Panels for Mechanically Stabilized Earth Walls

GEORGE HEARN, SCOTT MYERS, AND ROBERT K. BARRETT

Analysis, design, and testing of independent reinforced concrete facing panels for mechanically stabilized earth (MSE) walls are reported. Panels are intended for use as full-height facing for a variety of mechanical reinforcements for fills, including geotextiles, polymer geogrids, and steel mesh. Panels provide a forming surface and permanent facing for MSE walls, but are independent of the reinforced fill. Panels are attached to stable MSE constructions with flexible anchors that limit the earth pressures that can act on panels. Loads on panels are minimal, and panel size and appearance may be tailored to the requirements of individual projects and sites, offering options in construction and in appearance of the finished wall not previously available. Independent facing was tested in a prototype MSE wall using Ottawa sand fill reinforced by a nonwoven geotextile. In the test, flexible anchors performed as expected; earth pressures on panels were bounded by anchor yield loads; and, beyond an initial loading determined by anchor strength, earth pressures on panels did not increase with added surcharge. The basis for design of independent facing systems, methods for stress analysis of independent facing panels, an outline of construction procedures for MSE walls with independent facing, options for anchors and panels in independent facings, and a test of a prototype independent facing panel are presented.

Mechanically stabilized earth (MSE) walls are used in many applications in highway projects. Their economy and performance, and the increasing familiarity of highway engineers with the technology are combining to make MSE walls more accepted and more widely used. But greater acceptance brings demands for greater adaptability in MSE designs. For example, the aesthetics of a wall are often important. Block facings and stacked panel facings are attractive, but some projects may need walls with monolithic fronts not broken by horizontal joints. In such cases, full-height facing units are required.

For block facings and stacked panel facings, each facing unit is attached to a few (typically two) layers of fill reinforcement. Full-height facing panels used in a conventional MSE wall are attached to all reinforcement layers. For full-height facing panels fabricated in reinforced concrete, attachment to all reinforcing layers can result in significant stresses in the panel. The high stresses, in turn, lead to designs with relatively heavy panels.

High stresses in full-height facing panels result from a deformation demand. During construction, deformation occurs naturally as reinforcements in the fill are mobilized. Deformation-driven stresses can be avoided if facing is able to move. This is the concept of independent panel facing. In this paper, a design for independently anchored facing panels is presented. Independent facing systems use flexible anchors to accommodate wall deformations and thereby reduce earth pressures on panels. Independent facings are compatible with many types of earth reinforcements, including geotextiles, geogrids, and woven wire products. The performance of an independent facing system is demonstrated in load testing of a laboratory prototype.

FACING SYSTEMS

Facings for MSE walls protect fill reinforcements, anchor the tension in reinforcements, and contain the fill at the front of the wall. In anchoring tension and containing fill, facings are a structural design solution for the front boundary of the wall. The designs of block facing and panel facing systems are determined by these structural functions. The size and shape of facing units are adapted for simple, positive connection to fill reinforcement and for efficient construction. Wrapped-front geotextile walls use no units for facing but are still designed to anchor tensions and contain fill.

The comparison of block facing and wrapped-front facing reveals that the role of facing in MSE walls is a matter of design. Block facings, by design, perform all three roles of protection, anchoring, and containment. Wrapped fronts do not rely on facing units for anchoring and containment. The facings have different forms but equivalent functions. A rational approach to design of facing systems then is to identify the desired functions of the facing, to check that the facing is compatible with the load and deformation demands that will be placed on it, and to ensure that strength requirements of the MSE wall are satisfied.

The development of independent facing follows from a statement of function. First, to reduce the time required for a crane in MSE wall construction, it is desired that all facing panels be placed in a single operation not tied to the progress of the construction of the reinforced fill. The panels serve as a forming surface for the fill. Second, to achieve a monolithic appearance for walls, the elimination of horizontal joints in the facing is desired. Both requirements could be met by full-height panel facings.

Facing used as a front-forming surface for reinforced fill must be able to accommodate horizontal deformations as fill reinforcements are mobilized. The facing must have a mode of articulation to accommodate gradual, outward movement of the facing during construction. In block facings, articulation is the product of minor slips and rotations at joints.

Full-height panels have no joints and therefore no articulation in the manner of block facing. A second mode of articulation is available, however. Facing may tilt about its base. By tilting, facing can accommodate horizontal movement but will not conform to the reinforced fill. Because facing will not conform, the link between facing and fill must be flexible to preclude large restraining forces. This implies that a full-height facing panel should not be attached

G. Hearn and S. Myers, Department of Civil, Environmental, and Architectural Engineering, University of Colorado, Boulder, Colo. 80302. R. K. Barrett, Colorado Department of Transportation, 4201 E. Arkansas Avenue, Denver, Colo. 80222.

to fill reinforcements but instead should use flexible anchors that extend into the reinforced fill. Because panels are not attached to the fill reinforcements, facing is said to be independent of the reinforced fill. Tensions in reinforcements not anchored by facing must be anchored by other means such as a wrapped front. MSE walls with independent facing therefore comprise

- A stable reinforced fill, typically with a wrapped front;
- Independent facing allowed to tilt about its base but anchored to the reinforced fill; and
- Flexible connections between panels and the fill to limit restraining forces on the facing.

Proceeding from these, standard designs of reinforced concrete panels and deformable steel anchors for panels for walls 3.1, 4.6, and 6.1 m (10, 15, and 20 ft) high have been developed. The design examples presented in this paper are all reinforced concrete panels, although panels may be designed in other materials following the methods presented here.

INDEPENDENT FACING SYSTEMS

An MSE wall constructed with an independent facing is shown in Figure 1. This wall has full-height reinforced concrete panels tied to a reinforced fill with flexible steel anchors. Steel anchors are two-part loop bar anchors that accommodate vertical and horizontal deformation in the fill. Inelastic bending of the loop bars gives the two-part anchor an elastic or perfectly plastic tension response under increasing outward movement. Because the independent facing is not attached to fill reinforcements, the design of facing is effectively divorced from the design of the reinforced fill. The specific strength and deformation characteristics of a reinforced fill do not, within broad limits, influence the design of an independent facing system.

Structural Design of Panels for Facing

Facing panels are subject to earth pressures from the reinforced fill. Apart from loads in panels during handling and placement, earth pressures are the significant load demand on independent facing. The total thrust on independent facing is controlled by the yield load of anchors. Once the anchors reach their yield load, the facing panel

will tilt and will not accept higher pressures. For stable reinforced fills, deflections cease once the fill reinforcement is mobilized.

Independent facing panels are designed for moments and shears due to earth pressures. The thrust on facing panels is known from anchor yield loads, but the distribution of earth pressures is needed to compute section forces. Here, it is noted that pressure distributions assumed in design often do not match actual pressure distributions in MSE walls. Where pressures on facings have been measured by load cells or could be computed from tension force in fill reinforcements, it is observed that earth pressures may have a triangular distribution, or may show a peak value near the midheight of a wall, or may show low pressure at midheight with higher pressures at the top and bottom of the wall (1–7). Therefore, to establish a design basis for independent facings, it is necessary to consider pressure distributions that satisfy statics, that provide conservative estimates of section forces in facings, and that are reasonable in terms of both accepted design methods (8) and the pressure distribution observed in experiments.

Three forms of pressure diagram are considered: a triangular pressure distribution, a rectangular distribution, and a parabolic distribution (Figure 2). For each pressure distribution, bending moments in independent facing panels are computed. In the figure, facing panels are height H and width b and are secured by four anchors placed in pairs at distances $H/4$ and $3H/4$ from the bottom of the wall. The peak lateral earth pressure for each diagram P_{Max} is determined by the yield load A of the anchors for facing panels. The value of P_{Max} is computed by using a moment balance about the base of the panel. The maximum earth pressure depends on the anchor yield capacity only, not on properties of the fill. For this value of maximum earth pressure, a restraining force R at the base of the panel must be present to satisfy equilibrium of horizontal forces. For a triangular earth pressure distribution, it is found that

$$R = 2A \qquad (1)$$

And the maximum bending moment in an independent facing panel subject to a triangular earth pressure is

$$M_{Max} = 0.27AH \qquad (2)$$

Similar procedures computing P_{Max}, R, and M_{Max} are followed for rectangular and parabolic pressure distributions (Figure 2). A triangular pressure distribution leads to the highest estimate of bending moment in panels. The triangular pressure distribution is

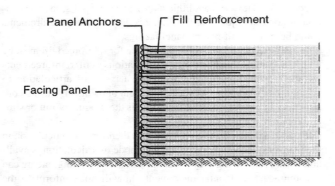

Section

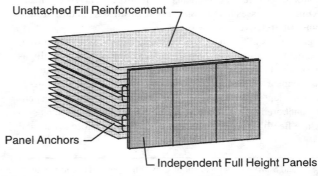

General View

FIGURE 1   Independent facing system.

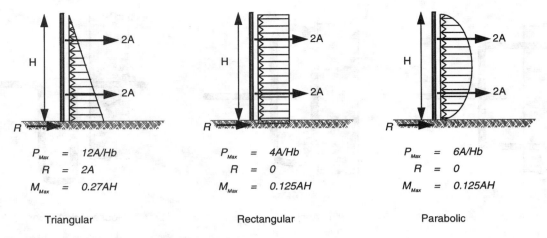

$$P_{Max} = 12A/Hb$$
$$R = 2A$$
$$M_{Max} = 0.27AH$$

Triangular

$$P_{Max} = 4A/Hb$$
$$R = 0$$
$$M_{Max} = 0.125AH$$

Rectangular

$$P_{Max} = 6A/Hb$$
$$R = 0$$
$$M_{Max} = 0.125AH$$

Parabolic

FIGURE 2   Trial soil pressure diagrams for design of facing panels.

adopted as a conservative design basis for facing panels in an independent system.

Taller panels use more anchors. For a vertical spacing of 1.5 m (5 ft) between anchors, panels at heights of 3.1, 4.6, and 6.1 m (10, 15, and 20 ft) use 4, 6, and 8 anchors, respectively. An increase in panel height corresponds to a fixed value of maximum earth pressure and an increase in maximum moment in panels. For all heights, moments and shears in facing panels are controlled by the yield load of the anchors. Results are shown in the "Statics" column of Table 1.

Structural Design of Anchors

Anchors for independent facing must allow movement of panels at moderate earth pressure, and must provide a permanent attachment of facing to the reinforced fill. The requirement for panel movement imposes an upper bound on anchor force that controls the earth pressures on facing panels. The need for permanent attachment of facing panels under self-weight, wind loads, and incidental loads imposes a lower bound on force in anchors. These two requirements may be met by anchors that yield at moderate load, that are capable of large movement during yielding, and that provide elastic response under external loading.

Three designs of anchors for panels have been developed (Figure 3). The first is a two-part design using a straight anchor bar in the reinforced fill attached to a loop bar on the facing panel. The straight anchor does not move; the loop bar provides articulation. The loop bar yields for outward tilt of facing panels. The vertical length of the loop bar allows the straight anchor to slip as fill settles. The loop bar may be bolted through a sleeve at the front of facing panels or may be attached to a plate at the vertical joint between panels. The bolted attachment allows an outward adjustment of panels that may

be needed to correct the alignment of facing panels after wall construction is complete.

Figure 3 also shows two other designs for flexible anchors. The blind anchor is a two-part anchor in which the loop bar is welded to a plate embedment in the facing panel. This design offers no adjustment of panel position. The gooseneck anchor is a one-part anchor. The neck in the anchor bar yields to allow outward movement of the facing panel. Gooseneck anchors have limited tolerance for vertical settlement of the reinforced fill.

The tensile load capacity of anchors is determined by the plastic bending strength of the loop bar or gooseneck. Considering the two-part anchors, the minimum yield capacity of the anchor can be computed as

$$A = 4M_p/l \qquad (3)$$

where M_p is the plastic bending capacity of the loop bar and l is its length. A two-part anchor will have its minimum strength when the straight anchor is located at the midheight of the loop bar. The anchor capacity will be higher when the straight portion is not at midheight. If the straight portion of the anchor is located at a distance l_a from the near end of the loop bar, the yield capacity of the two part anchor is

$$A = 2M_p\left(\frac{1}{l_a} + \frac{1}{l - l_a}\right) \qquad (4)$$

In service, anchors may not be located at midheight of loop bars due to settlement of the backfill, and due to normal construction tolerances. It is necessary to recognize two estimates of strength of flexible anchors. The minimum anchor strength is used for design against external loads on panels. A higher estimate of anchor load using an assumed attachment at $l_a = l/4$ is used to compute earth pressures and to design the facing panels.

TABLE 1   Statical Relations and Design Data for Panels

Panel Height (m)	Anchors (count)	Statics			Anchor Force (N)			P_{Max} (kPa)	Moments in Panels (N-m)				Panel Thick. (mm)	Rebars Gr 60 (Two Way)
		P_{Max} (kPa)	R (N)	M_{Max} (N-m)	Tilt A_{Stabl}	Wind A_{Wind}	Ult. A_u		Tilt M_{Stabl}	Wind M_{Wind}	Earth P. M_p	Ult. M_u		
3.1	4	3.9 A / b	2A	0.82A	56	2,670	3,540	5.7	-27	-1,020	2,920	3,240	127	#4@305 mm
4.6	6	3.9 A / b	3A	2.0A	67	2,670	3,560	5.7	-33	-1,020	7,300	8,110	152	#4@203 mm
6.1	8	3.9 A / b	4A	3.5A	67	2,670	3,560	5.7	-33	-1,020	12,800	14,200	152[a]	#4@203 mm

[a] Panel with two 254 mm deep webs.

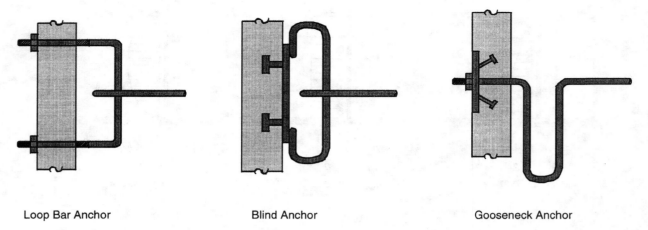

Loop Bar Anchor　　　　　Blind Anchor　　　　　Gooseneck Anchor

FIGURE 3　Flexible anchors for independent facing systems.

External load demand on anchors are wind and accidental eccentricity of panels. Wind load demand on a single anchor, A_{Wind}, is computed as

$$A_{Wind} = wbH/n \qquad (5)$$

where

w = design wind pressure,
b and H = panel width and height, and
n = number of anchors connected to the panel.

If panels are eccentric (tilted) and if the eccentricity is outward, then a force in the anchors A_{Stabl} is required to maintain stability of the facing. Figure 4 shows three conditions of panel eccentricity: a 3.1-m (10-ft) tall full-height panel tilted outward by an amount e, a 6.1-m (20-ft) tall full-height panel tilted outward by an amount e, and a 6.1-m (20-ft) tall stacked panel system displaced in the first tier. For full-height panels, the anchor force required for stability is computed as

3.1-m (10-ft) panel using four anchors

$$A_{Stabl} = \frac{We}{4H}$$

6.1-m (20-ft) panel using eight anchors

$$A_{Stabl} = \frac{We}{8H} \qquad (6)$$

where W is the dead weight of the facing panel. Using an estimate of e/H as 1/100, the anchor loads for stability can be expressed as

3.1-m (10-ft) panel using four anchors

$$A_{Stabl} = \frac{W}{400}$$

6.1-m (20-ft) panel using eight anchors

$$A_{Stabl} = \frac{W}{800} \qquad (7)$$

Design of anchors for independent facing proceeds by computing the required minimum anchor loads for wind and eccentricity loads and selecting an anchor with a yield capacity that exceeds these demands by an adequate margin of safety. In this study, the strength design provisions of the AASHTO specifications are followed (9). The yield capacity of anchors is then used to compute the earth pressures on facings. Example designs are presented in Table 1. The columns labeled "Anchor Force" show the load demands and design load for anchors. The wind load is taken as 1.4 kPa (30 psi), and panels are assumed to be normal weight concrete panels 2.4 m (8 ft) wide. Panels are 127 mm (5 in.) thick for 3.1 m (10 ft) height, and 152 mm (6 in.) thick for 4.6-m (15-ft) and for 6.1-m (20-ft) panels. Wind load controls the strength design of anchors. Table 1 lists bending moments in panels for tilt, for wind, and for anchor-controlled earth pressures. The table also lists rebar requirements for concrete panels. For panels, a concrete compres-

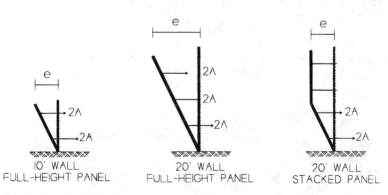

FIGURE 4　Stability of independent facing systems.

sive strength of 35 MPa (5,000 psi) and a rebar tensile strength of 413 MPa (60 ksi) are assumed.

Structural Design of Reinforced Fill

Independent facing panels are not attached to reinforcements in the fill, do not provide an anchorage for tensions in fill reinforcements, and offer only a limited capacity for retaining fill at the front of an MSE wall. MSE wall constructions may take advantage of facing panels as a forming surface during construction, but otherwise MSE walls using independent facing panels must be stable within themselves. Standard design procedures are available to ensure that MSE walls have adequate margins of safety against external failure mechanisms (i.e., sliding, bearing failure, and overturning) and against internal failure mechanisms, including rupture, pullout, and degradation of reinforcements. In addition, methods and analyses are available for designing MSE walls to satisfy limits on defections.

Construction of Independent Facing Systems

Construction of MSE walls with independent facing follows a sequence shown in Figure 5. Here, footings for panels are placed, and facing panels are moved into position and braced. Panels are keyed into footings, but there are no other attachments and no rebars across the joint. Bracing at the front of panels is removed when there are a sufficient number of anchors in place to support the facing.

Panel movement during construction may result in an unacceptable facing alignment. Two measures in construction offer remedies. At initial placement, facing units should be battered in

anticipation of a horizontal deformation. Inward batter on the order of 50 to 75 mm (2 to 3 in.) per 3.1 m (10 ft) of wall height is typical. After wall construction is complete, anchor connections may be loosened at the front of the wall and panels pulled forward if necessary to improve alignment.

Laboratory Demonstration of Independent Facing for MSE Walls

A full-height independent facing panel was used in the construction and load testing of two prototype walls in the laboratory. The prototypes were geotextile-reinforced walls approximately 3.1 m (10 ft) tall, 1.2 m (4 ft) wide, and 2.4 m (8 ft) deep. The prototypes each represent a slice of a wall of large lateral extent. The test fixture is a plexiglass box supported by steel strongbacks. It is equipped with greased membranes along the sidewalls to allow the fill to move with little side friction. Details of the test fixture are reported elsewhere (*10*). A general view of the prototypes is provided in Figure 6. The wall tests had two purposes: a demonstration of the performance of an independent facing system, and an investigation of the use of MSE walls with unwrapped reinforcement at the front. Fill reinforcements in these tests were neither attached to facing panels nor wrapped.

The two tests differed in fill material and in the sequence of loading. The first test used an Ottawa sand fill and the application of surcharge in several steps to a maximum of 138 kPa (20 psi). This test demonstrated the performance of independent facing and flexible anchors. The second test used a fill of Colorado DOT Class 1 road base. Surcharge was again applied in steps, but at each new loading the nuts restraining the flexible anchors were loosened and the wall was allowed to stand for a time. The repeated loosening of anchors was part of an effort to observe equilibrium in a fill with unwrapped,

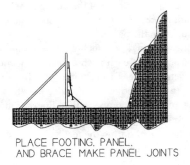

PLACE FOOTING, PANEL, AND BRACE MAKE PANEL JOINTS

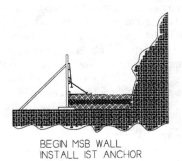

BEGIN MSB WALL INSTALL 1ST ANCHOR

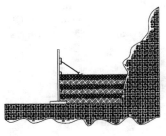

INSTALL SECOND ANCHOR REMOVE BRACE

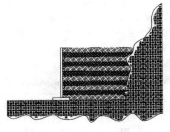

FINISH WALL

FIGURE 5   Construction sequence of independent facing systems.

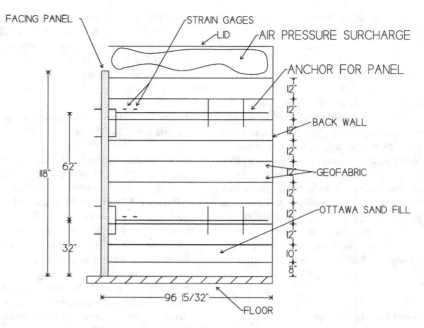

FIGURE 6 Prototype MSE wall with independent full-height facing.

unattached fill reinforcements. Only the first, Ottawa sand, test will be considered in this paper. Additional detail on the testing program can be found elsewhere (*11*).

Properties of fill reinforcements are listed in Table 2. The facing panel was a reinforced concrete panel approximately 3.1 × 1.2 × 102 mm (10 × 4 × 4 in.) with a two-way mat of #4 reinforcing bars at 127-mm (5-in.) spacing. The compressive strength of the concrete was 34 kPa (5,000 psi). Concrete reinforcing steel had a yield stress of 413 MPa (60 psi). The panel was provided with sleeves to accommodate adjustable loop-bar anchors. Loop bars were 13 mm (½ in.) in diameter and 305 mm (12 in.) long fabricated from smooth round bars. The straight anchor bars extended 2.1 m (7 ft) into the reinforced fill. Straight anchors were fitted with steel disks to improve pullout strength. Steel for anchors and loop bars had a yield strength of 289 MPa (42 ksi). Ottawa sand used for fill had a specific gravity of 2.65 and maximum and minimum unit weights per ASTM D-854 of 1 795 kg/m³ and 1 560 kg/m³ (112.2 pcf and 97.5 pcf) respectively. The sand reached a compacted density of 1 712 kg/m³ (107 pcf).

Loading on the wall was a surcharge made up of a 407-mm (16-in.) layer of sand and an additional air pressure applied at the top of the wall by a rubber bladder reacting against the lid of the test fixture. Loads applied by air pressure could be held constant over time to observe creep. The execution of loading on test walls included the application of air-pressure surcharge at 7-kPa (1-psi)

and 35-kPa (5-psi) increments, and the maintenance of surcharge. Loads were increased until some portion of the wall or the test setup failed. Failures included the seals around the panel and the air bag applying the surcharge.

Instrumentation for the tests included resistance strain gauges on all four anchors, six earth pressure cells mounted in the facing panel, resistance strain gauges on selected geotextile layers, dial gauges at five locations on the front surface of the facing panel, and a scribed grid on the sidewall membranes of the prototype. To monitor the performance of the facing panels and the anchors, the information needed is provided by strain gauges on anchors and by dial gauges on the panel.

Strain gauges on anchors were mounted in pairs on the straight-bar portion of each anchor near connections to loop bars. The pair of active gauges were wired in a full bridge with two additional gauges mounted on an unloaded length of steel round stock to serve as temperature compensation. For the Ottawa sand test, a single pair of strain gauges was mounted on each anchor. The gauges on one anchor failed during the test.

Four dial gauges were mounted at the corners of the facing panel and a fifth dial gauge was mounted at the middle of the top edge of the panel (Figure 7). From this pattern of gauges, it is possible to compute the translation, tilt, and twist of the panel.

In testing of the wall with Ottawa sand fill, air-pressure surcharges was applied at pressures of 7, 35, 69, and 138 kPa

TABLE 2   Properties of Geotextile Reinforcement for Prototype Test

Unit weight (ASTM D-3776)	1.93 N/m²
Grab tensile (ASTM D-4632)	890 N
Elongation at break (ASTM D-4632)	60 %
Modulus at 10 % elongation (ASTM D-4632)	4.45 KN/m
Coefficient of permeability	1.99*10-4 cm/sec
Nominal thickness	0.508 mm

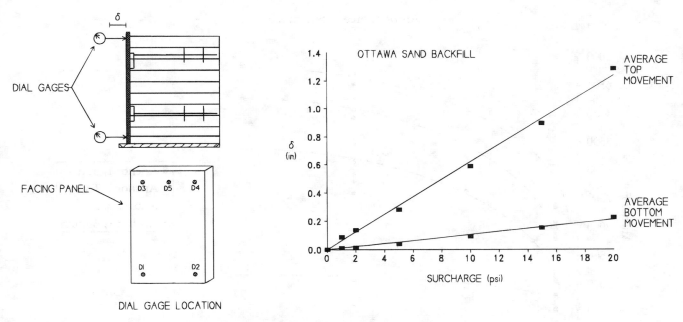

FIGURE 7   Dial gauge locations and deflection of panel versus surcharge.

(1, 5, 10, 15, and 20 psi). The test was stopped after the failure of a seal between the facing panel and the sidewall of the test fixture. The 7-kPa (1-psi) surcharge was held for approximately 75 hr. The 35-kPa (5-psi) surcharge was held for 30 min. The 69-kPa (10-psi) surcharge was held for 12 hr. The 103-kPa (15-psi) surcharge was held for 30 min. The 138-kPa (20-psi) surcharge was held for only a few minutes before a gasket at one vertical edge of the facing panel began to leak fill. The load history of the test is listed in Table 3.

The average movement at the top and at the bottom of the panel are plotted against surcharge in Figure 7. Loads in anchors are plotted versus surcharge in Figure 8. The anchor loads are determined directly from strain gauge readings. The strain gauges on Anchor No. 4 failed early in the test. From these figures several aspects of the performance of independent facing may be noted.

• Under surcharge, panel movement occurs by a combination of tilting and sliding. Panel deflection shows an essentially linear response to surcharge.

• Anchors exhibit a yielding response to increasing surcharge. Forces in two (of three) anchors show an upper bound load of about 3.6 kN (800 lb). The third anchor showed an upper bound load slightly greater than 4.5 kN (1,000 lb). All anchors exhibit greater stiffness initially, followed by a softening response at increasing

surcharge (Figure 8). This softening response is the intended yielding of anchors to limit earth pressures on facing panels.

• Anchor forces did not appear to vary with time at constant surcharge. However, two surcharge levels were maintained for periods of less than 1 hr. Long-term behavior of the wall with unwrapped reinforcement was not established in this test.

• Anchor forces exhibit a yielding response as a function of panel displacement (Figure 8). It is found that the anchor loads exhibit a softening behavior for the linearly increasing panel deflections. Again, this is the intended yielding behavior of anchors.

Analysis of Panels and Anchors in Prototype Tests

Following the procedures developed for design of panels, anchor loads are used to compute peak earth pressures for triangular pressure distributions at each level of surcharge. The results are plotted in Figure 9. Peak lateral earth pressures on panels are as high as 15 kPa (2.2 psi) for a surcharge of 138 kPa (20 psi). This peak pressure is substantially lower than the active earth pressure that would be computed for an MSE wall with reinforcements attached to facing. The lateral pressure on independent facing are not linear with surcharge. Moreover, lateral earth pressures are indeed bounded by the yield capacity of anchors for facing panels. The computation of

TABLE 3   Loading Sequence and Dial Gauge Readings for Test with Ottawa Sand Fill

Step	Time (hrs)	Action	Surcharge (kPa)	Dial 1 (mm)	Dial 2 (mm)	Dial 3 (mm)	Dial 4 (mm)	Dial 5 (mm)	Trans (mm)	Tilt (mm)
1	0	Wall completed	0	0	0	0	0	0	0	0
2	39.7	Applied 7 kPa	7	0.05	0.08	0.28	0.36	0.0	0.19	0.14
3	115.6	Additional 28 kPa	34	0.89	1.02	7.21	7.19	6.96	4.06	3.56
4	116	Additional 34 kPa	69	2.06	2.24	14.02	14.07	13.67	8.13	6.60
5	137.2	Additional 34 kPa psi	103	3.61	4.09	22.48	23.04	22.83	13.21	10.41
6	137.8	Additional 34 kPa	138	5.18	6.35	32.77	32.77	32.64	19.30	14.99

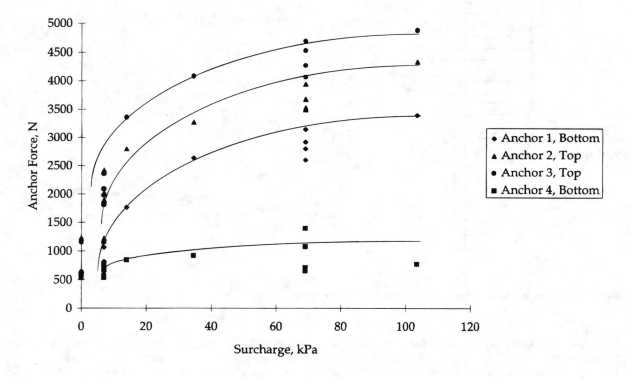

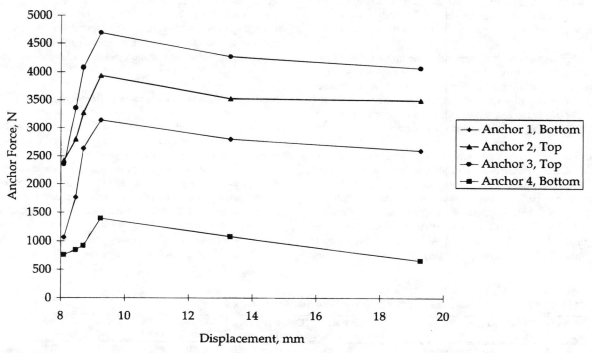

FIGURE 8 Anchor loads versus surcharge and versus displacement.

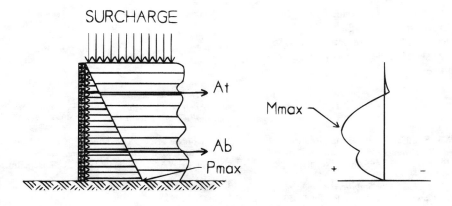

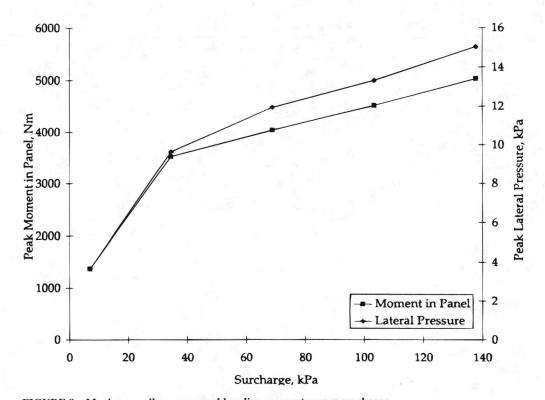

FIGURE 9   Maximum soil pressure and bending moment versus surcharge.

bending moments in panels follows directly from the computation of earth pressures. The highest bending moment in the panel is just over 60 kN-m (44,000 ft-lb) at a surcharge of 138 kPa (20 psi). Bending moments are also limited by the yield capacity of anchors.

CONCLUSION

Independent facing for MSE walls offers important options in design, construction, and aesthetics. Independent facing panels enjoy an articulation by a combination of sliding and tilting. Anchors for panels provide an upper bound load associated with the yield capacity of the loop bar. Once yielding is initiated, anchor

forces do not continue to increase with increasing surcharge or increasing panel movement. Yielding anchors impose an upper bound on the magnitude of lateral earth pressures acting on panels. Anchors are designed to provide adequate support of facing panels and at the same time to protect panels against high earth pressures. The design basis for independent facing computes maximum bending moments in panels as a function of panel dimensions and anchor yield load.

Independent facing and flexible anchors performed as expected in tests of prototype walls. It was observed that anchors yield smoothly with increasing surcharge and increasing displacement and that anchor loads reach a limiting yield load beyond which additional surcharge will not produce higher anchor forces. A 3.1-m-tall

(10-ft-tall) prototype wall with full-height independent facing was subject to an air pressure surcharge of 138 kPa (2,880 psf). At this surcharge, the maximum lateral earth pressure acting on facing panels was only 15.2 kPa (317 psf). Flexible anchors protected the facing from higher earth pressures.

REFERENCES

1. Anderson, L., K. Sharp, B. Woodward, and R. Windward. *Performance of the Rainier Avenue Welded Wire Retaining Wall, Seattle*. Department of Civil and Environmental Engineering, Utah State University, Logan, and CH2M Hill Engineers, Seattle, Wash., 1992.
2. Bressette, T., and G. Chang. *Mechanically Stabilized Embankments Constructed With Low Quality Backfill: Final Report*. Report T-633368 FHWA/CA/TL-86106. Caltrans, Sacramento, Calif., 1986.
3. Dondi, G. Load Test on a Retaining Wall Reinforced With Geosynthetics. *Proc., 4th International Conference on Geotextiles, Geomembranes, and Related Products*, The Hague, the Netherlands, 1990, pp. 101–106.
4. Hajiali, F. Field Behavior of a Reinforced Earth Wall in Malaysian Conditions. *Proc., International Reinforced Soil Conference*, Glasgow, Scotland, 1990, pp. 169–174.
5. Jakura, K. A., G. Chang, and D. Castanon. *Evaluation of Welded Wire Retaining Wall: Final Report*. Report 633370 FHWA. Caltrans, Sacramento, Calif., 1989.
6. Myers, S. *Independent Facing for Mechanically Stabilized Earth Retaining Walls*. Master's thesis, University of Colorado, Boulder, 1994.
7. *Tensar Geogrid-Reinforced Soil Wall: Grade Separation Structures on the Tanque Verde–Wrightstown–Pantano Roads Intersection, Tucson, Arizona, Experimental Project 1, Ground Modification Techniques: Final Report*. Report FHWA/EP-90-001-005. Desert Earth Engineering, Tucson, Ariz., 1989.
8. Christopher, B. R., S. A. Gill, I. P. Giroud, I. Juran, and J. K. Mitchell. *Reinforced Soil Structures, Volume 1: Design and Construction Guidelines*. FHWA/RD-89/043. FHWA, U.S. Department of Transportation, 1989.
9. *Standard Specifications for Highway Bridges*. AASHTO, Washington, D.C., 1992.
10. Wu, J. T. H. Predicting Performance of the Denver Walls: General Report. In *Geosynthetic-Reinforced Soil Retaining Walls* (J. T. H. Wu, ed.), Balkema, 1992, pp. 2–20.
11. Hearn, G., and S. Myers. *Independent Facing for Mechanically Stabilized Earth Retaining Walls*. CTI-CU-3-94. Colorado Transportation Institute, Denver, 1993.

Publication of this paper sponsored by Committee on Geosynthetics.

Biotechnical Stabilization of Steepened Slopes

DONALD H. GRAY AND ROBBIN B. SOTIR

The use of tensile inclusions makes it possible to repair slope failures or to construct steepened slopes along highway rights-of-way. Live cut brush layers can be used in place of or with synthetic fabrics or polymeric geogrids for this purpose. This approach, which is termed biotechnical stabilization or soil bioengineering, entails the use of living vegetation (primarily cut, woody plant material) that is purposely arranged and imbedded in the ground to prevent surficial erosion and to arrest shallow mass movement. In the case of brush layering, the live cut stems and branches provide immediate reinforcement; secondary stabilization occurs as a result of adventitious rooting along the length of buried stems. Imbedded brush layers also act as horizontal drains and wicks that favorably modify the hydrologic regime in the slope. The basic principles of biotechnical stabilization are described. Guidelines are presented for analyzing the surficial, internal, and global stability of brush layer–reinforced fills. A case study is reviewed in which live brush-layer inclusions were used to stabilize steep slopes along a roadway. A brush-layer buttress fill was used to repair an unstable cut slope along a highway in Massachusetts. Several repair alternatives were considered in this case. Scenic and environmental considerations with stability analyses eventually dictated the use of a composite, drained rock, and earthen brush-layer fill. The rock section was placed at the bottom to intercept critical failure surfaces that passed through the toe of the slope. Biotechnical stabilization resulted in a satisfactory and cost-effective solution; the treated slope has remained stable, and it blends in naturally with its surroundings.

Reinforced or mechanically stabilized earth (MSE) embankments have been used in highway construction for the past 2 decades. This approach offers several advantages over more traditional methods of grade separation that use either vertical walls or conventional fills with relatively flat slopes (2H:1V or less). The most prominent use of MSE is probably the widening and reconstruction of existing roads and highways. The use of reinforced steepened slopes to widen roadways improves mass stability, reduces fill requirements, eliminates additional rights-of-way, and often speeds construction. Design procedures, advantages, and several case histories of steepened, reinforced highway slopes can be found elsewhere (1).

The principal components of reinforced or mechanically stabilized earth embankments are shown schematically in Figure 1. Tensile inclusions (reinforcements) in the fill soil create a structurally stable composite mass. These main tensile elements are referred to as "primary" reinforcement. Shorter, intermediate inclusions may be placed near the slope face. These "secondary" reinforcing elements are used to minimize sloughing or face sliding and to aid compaction and alignment control. The soil at the outer edge of the slope may also be faced with some kind of netting (e.g., coir or jute) to prevent or minimize soil erosion. This last compo-

D. H. Gray, Department of Civil Engineering, The University of Michigan, Ann Arbor, Mich. 49109. R. B. Sotir, Robbin Sotir & Assoc., 434 Villa Rica Rd., Marietta, Ga. 30064.

nent can be eliminated, however, by simply wrapping the secondary reinforcement around the slope face of successive lifts or layers of soil as the embankment is raised. Stability considerations also dictate that appropriate external and internal drainage provisions be incorporated in the design.

Metallic strips, geotextiles, and polymer and wire grids have all been used as reinforcing elements in earthen slopes. Higher-strength, primary reinforcements are used for permanent, critical highway slopes. Lower-strength tensile inclusions can be used close to the face as secondary reinforcements. The latter are typically 0.92–1.8 m (3–6 ft) long and are spaced 203–914 mm (8–36 in.) vertically apart as shown in Figure 1. Selection of the appropriate reinforcement depends on the allowable tensile load, deformation, and design life of the structure.

The purpose of this paper is to describe the use of live cut brush layers as a supplement or alternative to inert tensile inclusions and to provide some guidelines for the design and installation of brush-layer reinforcements. The live brush can be substituted for the secondary reinforcements or, in some cases, actually replace both secondary and primary reinforcements. Unlike most inert reinforcements, imbedded brush layers also act as horizontal drains and wicks that favorably modify the hydrologic regime near the face of the slope. This approach, which is termed biotechnical stabilization or soil bioengineering, entails the use of living vegetation, primarily cut woody plant material, that is arranged and imbedded in the ground in selected patterns and arrays to prevent surficial erosion and to arrest shallow mass movement.

PRINCIPLES OF BIOTECHNICAL STABILIZATION

Live cut brush, woody stems, and roots can be used to create a stable, composite earth mass. The functional value of vegetation in this regard has now been well established (2). Biotechnical stabilization (3) refers to the integrated or combined use of living vegetation and inert structural. Soil bioengineering (4) is a more restrictive term that refers primarily to the use of live plants and plant parts alone. Live cuttings and stems are imbedded and arranged in the ground where they serve as soil reinforcements, horizontal drains, barriers to earth movement, and hydraulic pumps or wicks. Live plants and plant parts can be used alone or with geotextiles or geogrids. The live cut stems and branches provide immediate reinforcement; secondary stabilization occurs as a result of adventitious rooting that occurs along the length of buried stems. Techniques such as live staking, wattling (fascines), brush layering, and so forth, fall into this category. The U.S. Department of Agriculture, Soil Conservation Service (5) now includes in its *Engineering Field Manual* guidelines for the use and installation of these soil bioengineering methods.

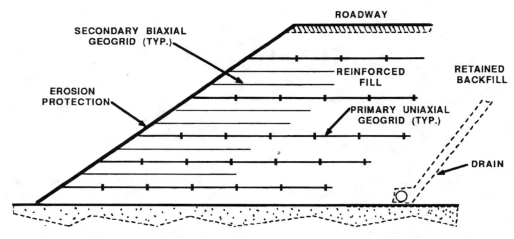

FIGURE 1 Material and structural components of a typical, reinforced steepened slope (*1*).

Brush layering consists of inserting live cut branches or brush between successive lifts or layers of compacted soil as shown in Figure 2. This process works best when done with the construction of a fill slope. The tips of the branches protrude just beyond the face of the fill where they intercept rainfall, slow runoff, and filter sediment out of the slope runoff. The stems of the branches extend back into the slope in much the same manner as conventional, inert reinforcements (e.g., geotextiles and geogrids) and act immediately as tensile inclusions or reinforcements. Unlike conventional reinforcements, however, the brush layers root along their lengths and also act as horizontal slope drains. This drainage function is very important and can greatly improve mass stability.

Brush layers alone will suffice to stabilize a slope where the main problem is surficial erosion or shallow face sliding. Sandy slopes with little or no cohesion fall into this category. Deeper-seated sliding tends to occur in embankment slopes composed of more fine-grained, cohesive soils. This situation may require the use of geogrids in combination with live brush layers. This latter approach is illustrated schematically in Figure 3. Guidelines are presented later in the paper for deciding whether geogrids must be used in conjunction with live brush layers.

BIOTECHNICAL STABILIZATION OF HIGHWAY CUT AND FILL SLOPES

Biotechnical stabilization has been used successfully to stabilize and repair steep slopes along highways. One of the earliest applications was reported in a work by Kraebel (*6*), who used contour wattling to stabilize steep fill slopes along the Angeles Crest highway in Southern California. Recent examples of soil bioengineering solutions for the stabilization of a highway cut slopes are found in

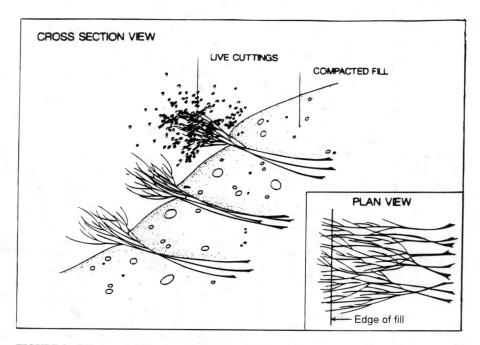

FIGURE 2 Fill slope stabilization using live brush layers place between lifts of compacted soil.

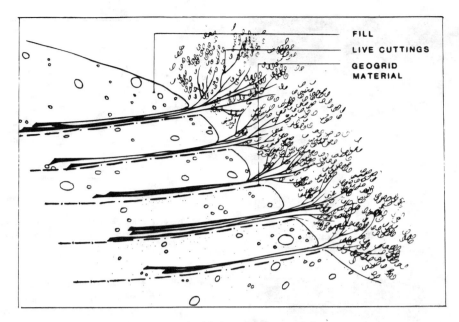

FIGURE 3   Live brush layers used with geogrids or geotextiles.

a work by Gray and Sotir (*7*). They also describe the use of brush-layering to repair a high, steep fill slope along a highway in North Carolina (*8*). An earthen brush-layer buttress fill was used to repair an unstable cut along a scenic highway in Massachusetts, as shown in Figures 4 and 5. The cut slope consisted of residual silty sand overlying fractured bedrock. Large amounts of groundwater seeped from fractures in the bedrock and through exposed soil in the cut. Other examples of brush-layer stabilization of a steep high-embankment slope along the Brenner Pass highway in Austria are shown in Figures 6 and 7.

STABILITY CONSIDERATIONS

Surficial Stability

One of the problems with embankment fills is the danger of erosion and sloughing along the outside edge of the fill. Several factors can

contribute to this problem, namely, poor compaction at the outside edge and loss of shear strength caused by moisture adsorption and low confining stresses. Attempts to improve compaction may be counterproductive because it impedes establishment of vegetation, which in the long run provides the best protection against erosion.

Brush layers are very effective in preventing shallow sliding and sloughing for the following reasons: (a) they act as wick and horizontal drains that intercept seepage and favorably modify the hydrologic regime; (b) they root along their length, and these adventitious roots provide secondary reinforcement or root cohesion near the slope face; (c) the growing tips of the brush layers slow and filter sediment from the slope runoff; and (d) the presence of the brush layers enhance the establishment of other vegetation on the slope face.

The effectiveness of mechanisms a and b can be demonstrated by "infinite slope" type analyses, which are appropriate for analyzing the surficial stability of slopes. For purposes of discussion consider a marginally stable, oversteepened (1.5H:1.0V) slope in a sandy soil, $\Phi = 35°$ and $\gamma = 118$ pcf (18.5 kN/m³), with very low cohe-

FIGURE 4   Brush-layer buttress fill immediately after construction (winter 1990, Greenfield Road, near Route 112, Colrain, Mass.).

FIGURE 5   Brush-layer buttress fill after 2 years showing extensive vegetative establishment (Greenfield Road, Colrain, Mass.).

FIGURE 6   Brush-layer embankment fill stabilization immediately after construction (Brenner Pass highway, Austria).

FIGURE 7   Brush-layer embankment fill stabilization after 2 years showing grass and brush establishment (Brenner Pass highway, Austria).

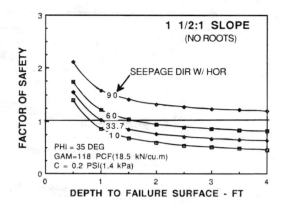

FIGURE 8   Factor of safety versus depth and seepage direction for 1.5:1 hypothetical slope without roots in the surface layer.

sion, $c = 0.2$ psi (1.4 kPa). Factors of safety can be computed as a function of vertical depth to the sliding surface (H) and seepage direction (θ) with respect to a horizontal reference plane as shown in Figure 8. In the absence of additional root cohesion, the factor of safety drops below unity ($F < 1$) when the seepage either parallels or emerges from the slope face at depths greater than 1 ft (0.3 m).

Brush layers and associated roots markedly improve surficial stability. The presence of fibers (roots) provides a measure of apparent cohesion *(9,10)*. This fiber or root cohesion can make a significant difference in the resistance to shallow sliding or shear displacement in sandy soils with little or no intrinsic cohesion. Actual shear tests in the laboratory and field *(9,11)* on root and fiber permeated sands indicate a shear strength increase per unit of fiber concentration ranging from 7.4 to 8.7 psi per pound of root per cubic foot of soil (3.2 to 3.7 kPa per kg of root/m³ of soil).

Root concentrations reported in actual field tests *(12,13)* were used to estimate likely root cohesion (c_R) as a function of depth. A low to medium root concentration with depth was used in the stability analyses to ascertain the likely influence of slope vegetation on mass stability. Factor of safety is shown plotted as a function of depth and seepage direction in the presence of root reinforcement for the same 1.5:1 slope in Figure 9. With roots present the safety factor is increased significantly near the surface and the critical sliding sur-

face is displaced downward. The results of the stability analyses show that both seepage direction (θ) and presence of root cohesion (c_R) have a significant effect on the factor of safety. Even a small amount of root cohesion can increase the factor of safety substantially near the surface. This influence is pronounced at shallow depths where root concentrations are highest and reinforcement effects therefore greatest.

The brush layers also act as horizontal drains and favorably modify the hydrologic regime near the face of the slope. They intercept groundwater flowing along the loose, outer edge of a compacted fill, divert the flow downward, and then convey it out laterally through the brush layer itself. Redirection of seepage flow downward in this manner results in greatly improved resistance to face sliding or sloughing *(14)*. Redirection of seepage from parallel flow direction ($\theta = 33°$) to vertical flow ($\theta = 90°$) greatly increases the factor of safety at all depths as shown in Figure 9.

In the case of highly erosive soils (fine sands and silty sands) and very steep slopes ($> 1.5H:1.0V$) it may be advisable to also use an erosion control netting or mat on the face of the slope between the brush layers. A biodegradable netting with relatively small apertures (e.g., coir netting) placed over long straw mulch will work well in this regard. The netting and mulch provide additional protection against erosion and promote establishment of vegetation on the slope face. The easiest way to install and secure the netting

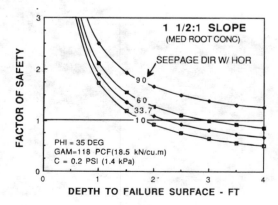

FIGURE 9   Factor of safety versus depth and seepage direction for 1.5:1 hypothetical slope with roots in the surface layer.

is by wrapping it around the outside edge of successive lifts of compacted fill.

Internal and Global Stability

The internal stability and global stability of a brush-layer fill slope protection system must also be considered. This is especially true when a brush-layer fill is used as a protective veneer or buttress fill against an unstable cut or natural slope. Sufficient tensile inclusions, either live brush layers or inert geogrids, or both, must be imbedded in the fill to resist the unbalanced lateral force acting on the earthen buttress. The brush stems and branches reinforce a fill in much the same manner as conventional polymeric grid or fabric reinforcements; accordingly, the internal stability of a brush-layer fill (i.e., the resistance of the brush reinforcement layers to pullout and tensile failure) can be analyzed using conventional methods developed for earth slopes reinforced with geotextiles or geogrids (*15,16*). The required vertical spacing and imbedded length of successive brush reinforcement layers are determined from the specified safety factor, allowable unit tensile strength, and interface friction properties of the reinforcement layer. The allowable unit tensile resistance for a brush layer can be calculated from the known tensile strength of the brush stems, their average diameter, and number of stems placed per unit width (*7*).

In the case of earthen fills that contain moderate amounts of low plasticity fines, the requirement for internal reinforcement is greatly reduced. The total required lateral resisting force approaches zero for fills with moderate cohesion (*c* = 300 psf or 14.3 kPa), slope inclinations less than 1.5H:1.0V, slope heights (H) less than 60 ft (18.3 m) as shown in Figure 10. Live brush layers used alone will suffice in this case to provide some additional internal stability, significantly increase surficial stability, and compensate for possible loss of intrinsic cohesion near the face. On the other hand, in the case of very high, steep slopes, a conservative design procedure would be to discount the influence of the live brush layers on internal stability and rely solely on the presence of inert tensile inclusions (e.g., geogrids, used in conjunction with the brush layers as shown in Figure 3).

Conventional geotechnical procedures can be used to analyze the global or deep-seated stability of brush-layer slope protection systems. A brush-layer reinforced outside edge of an embankment fill or alternatively a brush-layer reinforced buttress fill or veneer placed against an unstable cut or natural slope is simply treated as a coherent gravity mass that is part of the slope. An example from an actual case study will be used to demonstrate this analysis procedure.

CASE STUDY EXAMPLE

Project Site

The project site is located along Greenfield Road, just off State Route 112, in northern Massachusetts near the village of Colrain. Widening and improvement of this scenic road resulted in encroachment on an adjacent, unstable hillside, which triggered cut slope failures. The slope stratigraphy consisted of a residual soil, a silty sand, overlying a fractured quartz-mica schist bedrock. The cut was excavated back at a design slope angle of 1.5:1; the inclination of the natural slope above the cut was approximately 3:1. Cut slope heights varied in general from 20 to 60 ft (6.1 to 18.3 m). Slope fail-

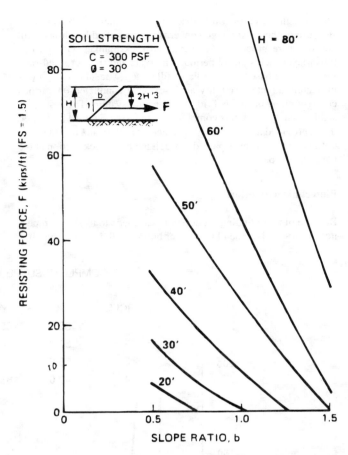

FIGURE 10 Chart solution for determining the required reinforcement or lateral resisting force for fills constructed from low-plasticity soils (*17*).

ures were characterized by small slipouts and slumping. A substantial amount of groundwater flowed out of the cut. This water seeped out of both fractures in the underlying bedrock and through the exposed face of the soil mantle.

Alternative Slope Treatments

The initial stabilization treatment of choice was a crushed rock blanket. This system is used frequently by Massachusetts Department of Transportation for cut slope stabilization. The main objection to this system was its stark and harsh appearance, which was inconsistent with the scenic nature of the highway. The main design consideration in the case of a rock blanket was to determine the thickness required to provide a specified global safety factor of 1.5. In fact, a crushed rock blanket placed the entire length of the slope was not required to satisfy mass stability. Instead, a drained rock buttress at the toe would have sufficed. A toe buttress, however, would have left upper portions of the slope exposed and vulnerable to piping and surficial erosion.

The soil bioengineering alternative proposed for the site was a drained brush-layer buttress fill. Reservations were expressed by the project engineer about the ability of an earthen brush-layer fill to resist large shear stresses at the base or toe of the slope and to provide a required global safety factor of 1.5. Some concern was also expressed about the possibility of a critical shear surface developing through the earthen fill adjacent and parallel to a brush layer. Because of these expressed concerns two modified brush-layer fill designs were proposed: (a) a crushed rock blanket with earthen brush-layer inclusions at periodic intervals and (b) a crushed-rock section at the base and brush-layer fill on top. The latter design was ultimately adopted; stability analyses were conducted on various configurations of this hybrid or composite system. The results of stability analyses on this composite system (see Figure 11) showed that it provided the required global factor of safety and that the most critical failure surfaces passed through the basal rock section at the toe of the slope.

Biotechnical Solution

Because of these findings, a decision was made to use the composite rock toe and earthen brush-layer buttress fill design to stabilize

the cut. An important caveat in this decision was the requirement that the earthen fill remain in a drained condition—a key assumption in the stability analyses. This requirement along with the large quantity of groundwater seeping out of the cut dictated that a suitable filter course or vertical drain be interposed between the earthen fill and cut face. This requirement was met by placing either a gravel filter course or a geotextile filter with adequate in-plane drainage capacity against the cut face during construction. Water from the bottom edge of the filter discharged into the rock toe at the base.

The construction work at the Colrain field site began in November 1989. A view of the cut slope after installation of the brush-layer buttress fill is shown in Figure 4. The appearance of the same slope some 2 years later is shown in Figure 5. In 2 years, the brush had fully leafed out and native vegetation had become well established on the slope. The slope is stable and has an attractive, natural appearance.

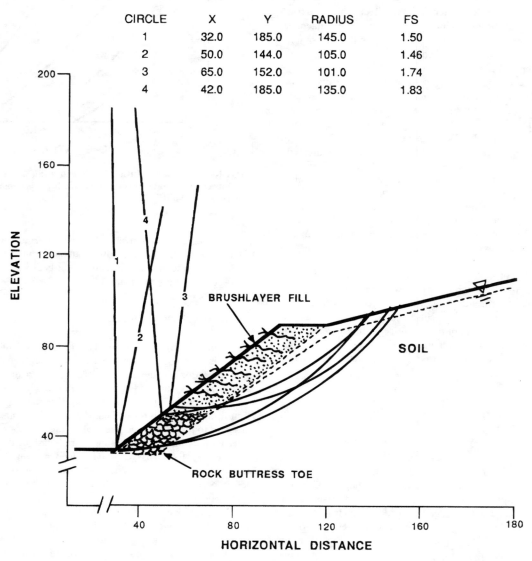

COMPLETE SLOPE CROSS SECTION

CIRCLE	X	Y	RADIUS	FS
1	32.0	185.0	145.0	1.50
2	50.0	144.0	105.0	1.46
3	65.0	152.0	101.0	1.74
4	42.0	185.0	135.0	1.83

FIGURE 11   Factor of safety calculated by Bishop Slope Stability analysis of cut slope stabilized by composite drained rock and earthen brush-layer fill (Colrain, Mass.).

Cost Analysis

The costs of several conventional slope stabilization treatments were determined and compared with the soil bioengineering treatment. The conventional treatment costs included a rock blanket and concrete crib wall. Cost analyses for the soil bioengineering treatment were conducted at two different stations or work locations on the project. The cost per square foot for the soil bioengineering treatment varied by only $2.90/m^2$ ($0.37/ft^2$) from one location to another.

The rock blanket costs included expenses for transporting, handling, and placing of 38 mm (1.5-in.) trap stone in a toe buttress or blanket 3 m (10 ft) high and 2.4 m (8 ft) wide. Placement of the rock higher up the slope entails greater difficulty and would have increased costs another 5 to 10 percent. The cost per square foot of front face for the crib wall includes footings and an estimated cost for the crib fill. The cost per square foot for the three alternative treatments was estimated as: rock blanket 2.5 m (8 ft) thick, $60.30/m^2$ ($5.60/ft^2$); soil bioengineering, $145.30/m^2$ ($13.50/ft^2$); concrete crib walls, $371.40/m^2$ ($34.50/ft^2$). Accordingly, the soil bioengineering costs were between those of a rock blanket and a concrete crib retaining wall. It should be kept in mind, however, that the contractor on the project had often placed rock blankets but had no previous experience with soil bioengineering. A cost comparison between these two methods was thus skewed slightly by unfamiliarity and a learning curve associated with the soil bioengineering method.

INSTALLATION GUIDELINES

Procedures for the harvesting, handling, storage, and installation of live plant material should be followed carefully. Successful biotechnical construction requires that harvesting and placement of live cuttings in the brush layers be carried out during the dormant season, usually November through April. Harvesting sites with suitable plant materials can be located with an aerial survey. Stems and branches up to 76 mm (3 in.) in diameter of willow, dogwood, alder, poplar, and viburnum shrubs are generally suitable for brush-layer treatments. They are cut at the harvesting site, bundled, and transported to the project site on covered flatbed or dump trucks.

Live cut material should be placed in the ground as soon after harvesting as possible. In the case of brush-layer installations, the cut stems and branches are laid atop successive lifts of compacted soil in a crisscross fashion (as shown schematically in Figure 2). Soil overlying each brush layer must be worked in between the branches to ensure contact between the brush and soil. The vertical spacing between brush layers normally varies from 0.30 to 0.91 m (1 to 3 ft) with closer spacings used at the bottom. The length of the cut stems should extend the full width, or as far as possible into an earthen buttress fill. A gravel drainage course, vertical chimney drains, or fabric filter with good in-plane drainage capacity must be placed between an earthen buttress fill and the cut face of a slope. Detailed guidelines and instructions for the selection, harvesting, handling, storage, and installation of live, cut plant materials can be found elsewhere (*5*).

CONCLUSION

Soil bioengineering solutions can be used to stabilize and repair slope failures along highway rights-of-way. Live brush layers can be used with or in place of inert polymeric reinforcements in oversteepened slopes. The growing tips of the brush layers filter soil from runoff and mitigate surficial erosion. The stems and adventitious roots in the brush layers reinforce the soil. The brush layers also act as horizontal drains and hydraulic wicks that favorably modify the hydrologic regime near the face of a slope. Stems and branches of plant species that root easily from cuttings (such as willow and alder) should be used. In addition, construction and installation should be carried out during the dormant season.

REFERENCES

1. Berg, R. R., R. P. Anderson, R. J. Race, and V. E. Curtis. Reinforced Soil Highway Slopes. In *Transportation Research Record 1288*, TRB, National Research Council, Washington, D.C., 1990, pp. 99–108.
2. Coppin, N., D. Barker, and I. Richards. *Use of Vegetation in Civil Engineering*. Butterworths Inc., Sevenoaks, Kent, England, 1990.
3. Gray, D. H., and A. T. Leiser. *Biotechnical Slope Protection and Erosion Control*. Van Nostrand Reinhold Inc., New York, 1982.
4. Schiechtl, H. *Bioengineering for Land Reclamation and Conservation*. University of Alberta Press, Edmonton, Canada, 1980.
5. Soil Bioengineering for Upland Slope Protection and Erosion Reduction. In *Engineering Field Handbook*, Part 650, 210-EFH, Soil Conservation Service, U.S. Department of Agriculture, 1993, 53 pp.
6. Kraebel, C. J. Erosion Control on Mountain Roads. *USDA Circular No. 380*, 1936, 43 pp.
7. Gray, D. H., and R. Sotir. Biotechnical Stabilization of a Highway Cut. *Journal of Geotechnical Engineering* (ASCE), Vol. 118, No. GT10, 1992, pp. 1395–1409.
8. Sotir, R. B., and D. H. Gray. Fill Slope Repair Using Soil Bioengineering Systems. *Proc., 20th International Erosion Control Association Conference*, Vancouver, B.C., Canada, 1989, pp. 413–429.
9. Gray, D. H., and H. Ohashi. Mechanics of Fiber Reinforcement in Sand. *Journal of Geotechnical Engineering* (ASCE), Vol. 112, No. GT3, 1983, pp. 335–353.
10. Maher, M., and D. H. Gray. Static Response of Sands Reinforced With Randomly Distributed Fibers. *Journal of Geotechnical Engineering* (ASCE), Vol. 116, No. 11, 1990, pp. 1661–1677.
11. Ziemer, R. R. Roots and Shallow Stability of Forested Slopes. *International Association of Hydrologic Sciences*, No. 132, 1981, pp. 343–361.
12. Riestenberg M. H., and S. Sovonick-Dunford. The Role of Woody Vegetation in Stabilizing Slopes in the Cincinnati Area. *Geologic Society of America Bulletin*, Vol. 94, 1983, pp. 504–518.
13. Shields, F. D., and D. H. Gray. Effects of Woody Vegetation on Sandy Levee Integrity. *Water Resources Bulletin*, Vol. 28, No. 5, 1992, pp. 917–931.
14. Cedergren, R. *Seepage, Drainage and Flownets*, 3rd ed. John Wiley & Sons, Inc., New York, 1989.
15. Leshchinsky, D., and A. J. Reinschmidt. Stability of Membrane Reinforced Slopes. *Journal of Geotechnical Engineering* (ASCE), Vol. 111, No. 11, 1985, pp. 1285–1300.
16. Thielen, D. L., and J. G. Collin. Geogrid Reinforcement for Surficial Stability of Slopes. *Proc., Geosynthetics '93 Conference*, Vancouver, B.C., Canada, 1993, pp. 229–244.
17. Lucia, P., and F. Callanan. Soil Reinforcement for Steeply Sloped Fills and Landslide Repairs. Presented at 66th Annual Meeting of the Transportation Research Board, 1987.

Publication of this paper sponsored by Committee on Geosynthetics.

Experiences with Mechanically Stabilized Structures and Native Soil Backfill

GORDON R. KELLER

Practices and experience with mechanically stabilized backfill retaining structures typically using native soil backfill on low and moderate standard rural roads by the U.S. Department of Agriculture, Forest Service, are documented. Information is provided describing innovative and low-cost alternative earth-reinforced retaining structures, including welded wire walls, chainlink fencing walls, geotextile walls, and walls faced with materials such as timbers, tires, hay bales, geocells, and concrete blocks. The design process has involved either generic or custom in-house designs, or proprietary designs with custom site adaptation and materials evaluation. Local, often marginal-quality backfill material is typically used. Its use is discussed, along with advantages and problems with marginal materials. Selected case histories with various wall types and backfill materials are presented.

The three basic objectives of this paper are to

• Document that the U.S. Department of Agriculture (USDA), Forest Service, has successfully constructed hundreds of mechanically stabilized backfill (MSB) structures nationwide over the past 20 years, typically using native soil backfill. These walls and reinforced fills, built with a wide variety of designs and construction materials, have performed well overall and satisfied their intended use.
• Discuss the Forest Service's retaining structure design process, and the merits and trade-offs of custom designs and use of in-house geotechnical personnel versus the use of commercial vendors and proprietary designs for structures.
• Document the successful use of local, often marginal backfill materials in most structures, and to discuss the advantages, disadvantages, and limitations of the use of marginal materials.

Considerable experience and knowledge have been gained in the use of relatively low-cost retaining structures for construction or repairs of rural roads with space constraints, particularly in steep mountainous terrain and unstable ground. Site access is often difficult and locations are remote, making the use of geosynthetics and soil reinforcement concepts, modular or prefabricated components, and on-site backfill materials highly desirable.

Composite facing and reinforcement elements used with on-site backfill material offer substantial cost and construction advantages over many conventional retaining structures. Simple construction techniques are desired and often necessary. Minimizing cost is often an objective. MSB structures discussed here are ideal for forest or rural applications as well as far many private, local, and public road and highway needs.

A wide variety of retaining structures has been used. Wall types, typically up to 7.5 m (25 ft) high, have including welded wire walls,

geotextile walls, chainlink fencing walls, lightweight sawdust walls, and walls faced with segmental concrete blocks, hay bales, tires, geocells, or timbers. Some soil-reinforced rigid concrete face panel structures have also been used. Reinforced fills with local embankment material have been an economical alternative to walls in some areas. Considerations for each of these types of structures are briefly discussed. Selected case histories that represent a range of structures and backfill materials used are presented.

Many walls are designed in house by geotechnical personnel using available design methodologies to take advantage of custom designs, risk assessment, and cost savings of earth reinforcement systems and local materials. Other walls are designed and constructed using readily available manufacturers' standard designs, along with laboratory testing to ensure that backfill material meets design parameters. Drainage is nearly always incorporated into designs, commonly with geocomposite drains. Filtration, durability, and transmissibility requirements for the geocomposite drainage systems are specified.

Local backfill material is most often used on Forest Service projects. Fortunately typical soils found in a mountain environment have a high friction angle, satisfying needed design strength criteria. However, fine-grained native soils can present design and construction problems, such as unacceptable deformation, poor compaction and drainage, and some risk. Nevertheless they may offer significant cost savings over conventional coarse granular backfill. Fine silty sands to silts with some clay and soils with up to 50 percent fines have successfully been used as backfill.

DESIGN PROCESS

One of three basic design approaches is used on Forest Service projects.

1. Custom retaining structures are selected, designed and constructed, or contracted by the Forest Service with technical input from in-house geotechnical personnel;
2. Vendor-provided structures and designs are selected by the Forest Service, with technical input from geotechnical personnel on wall type, loading conditions, foundation and site evaluation, and so forth or;
3. A consultant-, contractor-, or vendor-provided design, with some site evaluation, is used with the approval of the Forest Service. Geotechnical personnel may or may not be involved in the process.

Most structures built have used either the first or second approach. Basic retaining structure selection and design information have been documented in the Forest Service *Retaining Wall Design Guide* (1).

U.S. Department of Agriculture, Forest Service, Plumas National Forest, P.O. Box 11,500, Quincy, Calif 95971.

The design process, type and thoroughness of site evaluation, and wall selection usually depend on the skills of the personnel involved in the project. Geotechnical personnel are not common in the Forest Service agency. Most regional offices have either an individual or small staff of geotechnical personnel. A few individual forests in the West have a staff geotechnical engineer or an engineer who is responsible for several forests. In any event, the geotechnical personnel are involved in a wide range of projects and are spread thin, and time and project involvement are always limited. Thus the time committed to any project depends on the current workload and priorities, and available time may dictate what type of retaining structure and design process to use.

The main advantages of custom in-house designs with unique structures include

- The ability to evaluate the full range of available structures,
- The ability to use local or surplus construction materials,
- The ability to realize the maximum cost savings, and
- An opportunity to advance the professional state-of-the-design practice, combining practical application with research and development.

Additional advantages of having geotechnical personnel knowledgeable of soil reinforcement concepts involved include the following:

- Staff has the opportunity to perform all aspects of the project, including site investigation, foundation assessment, materials evaluation, construction control, drainage needs, and external and global stability analysis, as well as overall design and details.
- Design and construction field changed conditions can be better evaluated and accommodated.
- The risk and trade-offs of various types of structures and materials used can be better assessed.
- Current developments by other agencies and within the profession can be used and implemented.
- Proper limitations and applications of earth reinforcement concepts can be made, and misuse avoided.

The following are the advantages of using vendor products or manufacturer's standard designs:

- Standard designs and trial solutions can be evaluated quickly.
- Good construction support is likely, which commonly goes along with use of manufacturers' products.
- With limited time and resources, internal design is satisfied, though perhaps conservatively, so available time can be spent on external and global stability, foundation conditions, and other project aspects.

In reality, the use of vendor-supplied designs and products has been satisfactory and necessary at times and has cost the agency only a limited amount of money. The differential construction cost of a vendor's wall versus custom-designed walls has typically been about $30 to $50 per square meter (a few dollars per square foot) of face. However, the minimum cost of a vendor-provided wall has been around $180/m^2 ($17/ft^2), and minimum in-house designed walls (geotextile walls) have cost $110/m^2 ($10/ft^2) of wall face. Still the major advantages of having in-house expertise are overall cost-effectiveness, the total evaluation that can be accomplished, and the flexibility it offers.

The actual design process used by agency geotechnical personnel has depended on time, information available, and type of wall desired. Most early geotextile-reinforced and chainlink fencing wall designs were based on the ultimate strength design method developed by the Forest Service (*2*). Welded wire walls were designed by or followed design tables developed by the Hilfiker Company or now use design information such as that presented in NCHRP 290 (*3*). Reinforced fill designs have used methods involving modified slope stability limit equilibrium analysis (*4*).

Today many design procedures are being proposed, refining the earlier relatively conservative design methods. A recommended synthesis of design procedures has been presented by FHWA (*5*). Also, generic and product specific PC based computer programs are available to facilitate the design process. For low- to medium-height structures, the standard designs available from manufacturers or the generic designs for low geotextile walls with given backfill and loading conditions (*6*) are very simple to use and practical in many applications. Note, however, that many manufacturer's PC programs are product specific and do not allow the user to check calculations independently.

MSB STRUCTURES

Many recent innovative designs have been developed using soil reinforcement concepts, and numerous walls have been built on rural roads using a variety of reinforcing, facing, and backfill materials. Of the walls constructed by the Forest Service in the past decade, MSB structures have been used at least 80 percent of the time, mainly because of cost and ease of construction. Most use local or on-site backfill material and easily fabricated flexible reinforcement elements. For walls less than 7.5 m (25 ft) high, cost has typically ranged from $160 to $270/m^2 ($15 to $25/ft^2) of face. Both frictional reinforcement systems (i.e., geotextiles) and passive resistance reinforcement systems (i.e., welded wire and geogrids) are commonly used.

Walls are often located on landslides or on sites with minimal foundation information, so some limited wall deformation is desirable. Site and foundation investigations are rare for small walls. Soil-reinforced structures that minimize foundation pressures, have relatively wide foundations, and tolerate deformation are desirable. Brief descriptions of many of the MSB structures used by the Forest Service follow.

For wall drainage, geocomposite drains have been successfully and extensively used since 1975. They are particularly applicable where the excavated back slope is steep or nearly vertical, making conventional gravel drains difficult to construct. Geocomposite drains on several wall sites in California, instrumented since 1984, have performed very well. Results reported elsewhere (*7,8*) show that many geocomposite drains available today have good crushing strength properties and adequate flow capacity and satisfy needed filtration criteria. However, available products performance varies considerably. The drains themselves cost $20 to $45/m^2 ($2 to $4/ft^2), installed.

Welded Wire Walls

Welded wire walls up to 9 m (30 ft) high are the most commonly used MSB system in the Forest Service (Figure 1). These walls have also been constructed to heights greater than 27 m (90 ft). Many

FIGURE 1 Seven-year-old welded wire wall with some face settlement due to use of native soil backfill (Plumas National Forest, California).

contractors are familiar with their assembly and local manufacturers' representatives have provided excellent construction support services to contractors, with technical advice and by providing on-the-job training to contractors. They are relatively easy to construct on grades and curves, can be adapted to many sites, can have a 50- to 75-year design life, and have often been used for bridge abutments and around culverts. Again, design information is readily available (3).

Use of fine backfill material in welded wire walls has occasionally led to vertical face settlement, from poor compaction along the wall face and from fine soil migration through the wire face of the wall. A layer of heavy ultraviolet-resistant geotextile is now usually placed against the wire mesh to contain the fine soil near the face. Use of tamped pea gravel or coarse material in the face zone will further minimize this problem and is generally recommended.

Limited experience has been gained on projects with welded wire soil reinforcement and use of rigid precast concrete face panels. The panels form a durable aesthetic wall facing. However the rigid panels are sensitive to foundation settlement and any face deformation. Select backfill is recommended with this facing system.

Geotextile Walls

Geotextile-reinforced soil walls were pioneered by the Forest Service in 1974. They have perhaps the least expensive materials cost of any wall available. For reinforcement materials alone, the cost is as low as $15/m² ($1.25/ft²) of wall face. Design procedures are widely published (1–3). These procedures have led to many successful and perhaps conservative designs for walls 3 to 6 m (10 to 20 ft) high or higher.

Reinforcement lift thickness (compacted) typically varies from 15 to 45 cm (6 to 18 in.). Thicker lifts are difficult to form. Base geotextile embedment length is typically nominal for pullout resistance and is dictated by the length required to resist sliding failure in external stability calculations.

Because of the flexibility of geotextiles, temporary forms must be used to support the wall face as each lift is constructed, making this process somewhat slow and labor intensive. Long 5- by 30-cm (2 by 12-in.) lumber and metal brackets are usually used for the forms.

Most geotextiles must be protected from long-term degradation by sunlight. A gunite layer is often applied to the wall face. In remote forest applications, a protective coating of asphalt emulsion may be specified, which must be repeated several times during its service life. The final wall face itself usually has an irregular shape, but its appearance is acceptable in most rural settings. This type of wall is also ideal for temporary construction applications.

Geogrid walls have been built using a design concept very similar to that used for geotextile walls, where additional strength and less creep are desired for a high wall or a stronger, more durable facing is wanted.

Lightweight Geotextile Walls

Several walls up to 8.5 m (28 ft) high have been constructed in Oregon and Washington using wood chips or sawdust. This material, wrapped in a geotextile, produced a lightweight structure ideal for placement on an active slide deposit. Design and construction procedures for this wall were roughly similar to those of a normal geotextile wall since wood chips have a high friction angle (25 to 40 degrees, based on triaxial tests). Wood chips were spread and compacted in 45-cm (18-in.) lifts between the reinforcing layers. Compaction was difficult to measure so a procedural specification of several passes per lift was used. A final typical moist density of approximately 6.3 kN/m³ (40 pcf) was achieved.

Gradation of the wood backfill used has ranged from fairly clean 75-mm (3-in.) maximum size chips to a fairly dirty sawdust. Performance has been satisfactory and settlement of the material after 10 years has been limited to about 5 percent of the structure height as slow decomposition continues.

Chainlink Fencing Walls

Several chainlink fencing walls up to 6.7 m (22 ft) high have been constructed by the Forest Service using conventional 9-gauge galvanized chainlink fencing material placed in 30- to 60-cm (12- to 24-in.) lifts in the backfill material for reinforcing.

Pullout resistance and strength parameters for custom design of the chainlink wall are similar to those of a welded wire material. The construction procedures for forming the face are similar to those used for a geotextile wall. A 6-mm (¼-in.) galvanized screen is placed at the wall face to confine the backfill material. Hay bales have also been used to form temporarily the face of this type of wall.

Timber-Faced Walls

An ideal type of reinforced soil wall appears to be one incorporating the ease and cost savings of geotextiles or geogrid reinforcement with durable and aesthetic timber or other facing members. A geotextile-timber wall developed in Colorado (9) appears to be a nearly ideal combination of materials (geotextile and railroad ties) that is easy to construct, aesthetic, and cost-effective. Several such walls, up to 5.5 m (18 ft) high, have been constructed by the Forest Service and appear pleasingly rustic and particularly appropriate for a rural or forest setting (Figure 2). However, because timbers are treated with wood preservatives, this type of wall facing system is not used near water courses.

The connection detail of the reinforcing material to the timbers varies. Techniques have included sandwiching the geotextile or

FIGURE 2 Timber-faced geogrid-reinforced wall in forest setting constructed with native soil backfill (San Juan National Forest, Colorado).

geogrid between the timbers, stapling the material to the timbers, wrapping it around the timbers and adding a face plate, and using an intermediate wrapped board sandwiched between the main timbers. Each facing connection technique has proven to be adequately strong for low to moderate-height walls given the relatively low lateral stress on the wall face with frequently spaced, extensible reinforcing materials (*10*). Timbers are often pinned together with spikes or rebar.

Segmental Concrete Block-Faced Walls

A wide variety of concrete block facings exists today; they are typically used with geogrid reinforcement. This combination of materials is easy to construct, cost-effective, and aesthetic, particularly in rocky areas. The walls have only recently been built by the Forest Service, using manufacturer-provided designs.

The connection detail of holding the geogrid in place with dowels set into the blocks is simple, easy to construct, and generally effective. Other face-connection techniques involve sandwiching the reinforcement between the blocks or shear keys in the blocks. On relatively high walls with native backfill, a drain should be located behind the backfill and behind the concrete blocks as recommended by manufacturers. State-of-the-practice information on use and limitations of segmental concrete block walls is presented elsewhere (*11*).

Tire-Faced Walls

Several tire-faced, earth-reinforced walls, up to 3.1 m (10 ft) high, have been designed and constructed by the Forest Service in northern California, using a slit-film woven geotextile reinforcement with used tires for the facing members. Because used tires are plentiful and free, geotextile is inexpensive; because construction is simple, the cost of this type of low wall is minimal. This type of wall is particularly easy to build with in-house crews or local hand labor, with minimal construction equipment, and is ideal for applications such as road shoulder support.

The design consists of layers of geotextile on a 19- to 38-cm (8- to 15-in.) vertical spacing placed between every one or two rows of tires. Tires are staggered on top of each other. Soil is compacted behind each layer of tires in 18- to 20-cm (7- to 8-in.) lifts. Local material is backfilled into each tire and hand compacted, filling effectively only the middle "hole" in the tire (*12*).

This type of wall needs to be built with a nearly 1H:4V face batter and tires staggered horizontally, one-half tire diameter on each successive layer, to prevent the backfill soil from falling through the hole and space between tires on the next lower layer. The stagger and vertical offset of the tires can provide planting space in the tire holes for vegetation, adding long-term biotechnical stabilization to the wall and improving appearance.

Settlement of the tire face is a limitation for this type of wall. After more than 5 years of monitoring one wall, deformation appears acceptable, with face settlement on the top row of tires of about 0.3 m (1 ft), or 10 percent of the wall height. Soil and pockets of vegetation have partially masked the tires.

Geocell-Faced Walls

Several geocell-type walls have been constructed by the Forest Service up to 6 m (20 ft) high since 1988. Walls have been either gravity structures or "zoned" gravity (geocell-reinforced) structures where some geocell layers extend into the backfill. These, geogrid and geotextile-reinforced geocell-faced wall designs, and testing information are discussed elsewhere (*13*).

The geocell fill and backfill materials have been native granular material. The cells provide confinement for loose, granular soils. The lightweight expandable cells are ideal for moving the material to remote sites. With a battered wall face, each "cell" forms a planter for vegetation. The dark or tan high-density polyethylene material has reasonable resistance to deterioration.

Reinforced Fills

Reinforced fills placed with a 1H:1V or steeper face slope have offered an economical alternative to retaining structures for those sites where the ground is too steep for a conventional fill slope yet is flat enough for a reinforced fill. Reinforced fill heights have ranged from 5 to 15 m (15 to 50 ft) on forest projects, and over twice this height elsewhere.

USE OF LOCAL BACKFILL MATERIAL

On-site or local materials, often of marginal quality, are consistently and successfully used by the Forest Service for backfill in retaining structures and reinforced embankments. They are desirable because of the unavailability or expense of imported materials. "Marginal" soils are defined as fine-grained, low-plasticity materials that may be difficult to compact, have poor drainage, or have strength parameters sensitive to density.

Coarse rock fill material, occasionally available, is excellent for backfill if it is well graded. Material with a 15-cm (6-in.) maximum size is commonly specified. However, rock fill often has enough oversize material to make layer placement difficult and to damage the reinforcement material. Free-draining rock fill is necessary only in special applications such as in coastal or streamside structures subject to periodic inundation.

Select granular free-draining backfill material, commonly specified by wall manufacturers, can be expensive to import. The cost advantage of using local material or material excavated on site can be significant, particularly in rural areas. The average cost of local backfill material, reflecting materials, placement, compaction, and haul cost, is estimated at \$10.50/m³ (\$8/yd³), compared with imported select backfill at roughly \$23.50/m³ (\$18/yd³). Given a medium-sized structure with 140 m² (1,500 ft²) of wall face and an estimated backfill quantity of 575 m³ (750 yd³) the differential cost, or savings, is \$7,500. Some of the savings may be offset by increased construction costs. However, nationwide, hundreds of thousands of dollars could be saved annually by using local backfill materials.

Select Material

Relatively clean free-draining granular soils are generally recommended by manufacturers and preferred by contractors as backfill for retaining structures. Select backfill material requirements recommended by AASHTO (14) for reinforced structures conform to the following gradation limits:

Sieve Size	Percentage Passing
10 cm (4 in.)	100
No. 40	0–60
No. 200	0–15

The plasticity index (PI) as determined by AASHTO T-90 should not exceed 6. The material should exhibit an angle of internal friction of not less than 34 degrees, at a compacted density of 95 percent of AASHTO T-99. No testing is needed for backfill where 80 percent of the material is larger than 1.9 cm (¾ in.). The material should be free of organic matter or other deleterious materials. Requirements also include other durability and corrosion considerations. Note that some agencies recommend use of material 1.9 cm (¾ in.) or smaller to prevent geosynthetic reinforcement damage.

Some manufacturers, such as the Reinforced Earth Company (Terre Armee Int.), will occasionally allow use of "intermediate" soils containing up to 40 percent fines, provided that the PI does not exceed 30 (15). Use of intermediate soils is limited to special cases, particularly outside the United States, where select materials are not available. With intermediate soils, specific design, evaluation, and careful construction control are required. Select material specifications are typically used, particularly where deformation must be minimized and where rigid face panels are used.

Marginal Backfill Material

Under many circumstances essentially any nonplastic to moderate plasticity, frictional soil can be used as backfill, provided the wall is designed to resist the external and internal forces. In remote areas it is generally more economical to use local native or fine-grained backfill, with drainage, and design for those appropriate strength parameters than to import select free-draining materials.

When fine-grained or marginal material is used as a wall backfill, several factors must be considered:

- The structure should be specifically designed for the strength properties of that material.
- Care should be taken to closely control placement moisture content and density.

- A well-designed and thorough drainage system should be included.
- The likelihood of accelerated corrosion must be evaluated.
- Relatively slow construction is likely, and slight to moderate formation and face settlement should be expected.

Known or documented failures (16) of soil reinforced walls with marginal or clay backfill material (which incidentally are uncommon) have ignored or overlooked one of these factors.

Local "marginal" materials used by the Forest Service have varied from silty sands to silts and clays [SM, SC, ML, and CL (Unified Soil Classification)] with over 50 percent fines (passing the No. 200 sieve) and a PI of up to 15. Marginal silt and clayey soils have been successfully used and evaluated by FHWA in its full-scale tests on the behavior of reinforced soil (10) and in the Denver test walls by the University of Colorado and the Colorado Department of Transportation (17).

Marginal materials should be specifically tested to determine their strength properties and strength-density relationship. Peak shear strength parameters should be used for analysis. Experience with marginal backfill has been favorable, but use can present problems in construction and long-term performance. Compaction of fine-grained soils is sensitive to moisture content, so close construction control is needed and the specified densities (typically 95 percent of AASHTO T-99) may be difficult to achieve. Construction delays may occur. However, once compaction is achieved, results have generally been satisfactory. Frost heave in cold regions can also be a problem under some circumstances.

Inability to achieve the specified compaction near the wall face, loss of fines through the face, or soil compressibility have resulted in some face settlement of structures. Most measured settlement has occurred in the first 2 years after construction, with minimal additional long-term settlement. Measured settlements have been 2 to 4 percent of the wall height.

Surface drainage should be designed to keep water from infiltrating into the backfill. With fine-grained low-permeability backfills, long-term saturation of the fill in a wet environment is possible, even with a drain installed behind the structure. To prevent saturation, layers of free-draining gravel can be built into the backfill. (However, this will add to the wall cost.) To prevent surface water from entering the fill, the backfill material may be waterproofed with a paved roadway surfacing.

Table 1 shows examples of local marginal soils that have been used successfully in Forest Service structures. Note that the soils, though fine grained, have good frictional characteristics and were compacted.

In view of the existing information and overall good performance of many MSB structures using marginal, fine-grained, low- to moderate-plasticity soil, it appears that industry standards could be modified to reflect this information and to realize the economic benefit of their use. Existing standards could include select material specifications suitable for high-risk and high structures or those with little tolerance to deformation or differential settlement. Most structures could be constructed using a standard (intermediate) backfill material, with limits such as 50 percent fines, a PI of 20 or less, and a minimum peak effective friction angle of 25 to 30 degrees.

Clay-Rich Cohesive Backfill Material

Generally poor-quality, clay-rich, cohesive soils with low frictional strength should not be used in retaining structures. Exceptions may

TABLE 1 Local Marginal Materials used in Forest Service Structures

Site & Forest	Wall Type (Height, m.)	USC	% Minus 200	PI	Phi' deg	c' kPa	Comments
Goat Hill Plumas NF	Welded Wire (4.6 m.)	SM	21	5	34	9.6	4% Settlement on Face, Most in 2 of the 10 Lifts.
		SC	20	8	31	14.4	
		SM	23	4	27	16.7	
Mosquito R. Tahoe NF	Welded Wire (8.2 m.)	SM	22	NP	-	-	Minor Settle- ment, Vegetated.
		ML	50	6			
L.North Fork Plumas NF	Reinforced Fill (1:1) (15.2 m.)	SM	38	2	34	4.8	Minor Slumping, Well Vegetated.
		ML	55	3	33	7.2	
Gallatin Lassen NF	HSE-Concrete Face/W. Wire (3.8 m.)	GW	1+	NP	30+	-	Minor Face Panel Separation using Light Cinder Fill
B.Longville Plumas NF	Welded Wire (5.5 m.)	CL- SM	50+	-	26	9.6	Poor Foundation, 3 % Settlement.
Grave Plumas NF	Geotextile (2.7 m.)	SM	26	NP	35	40.7	Irregular Face, Weathering Cloth but no Fill Loss.
		SM	15	NP	38	26.3	
Butt Valley Plumas NF	Tire-Faced (3.1 m.)	SC	38	8	26	19.2	10% Face Settlement.
Thomjack Klamath NF	Timber-Faced (4.6 m.)	SM	27	NP	30+	0	Minimal Settlement.
Stump Spring Sierra NF	Welded Wire (6.8 m.)	SM- SC	42	15	-	-	Performing Well, Min. Settlement.
Pulga Plumas NF	Welded Wire (5.9 m.)	SM- GM	44	4	29	9.6	Mod. Settlement, Poor Compaction.
Agness Siskiyou NF	Chainlink Fencing (to 6.7 m.)	GM- SM	15	NP	-	-	Min. Settlement, Min. Corrosion, Face Vegetated.
Camp 5 Hill Willamette NF	Wood Chips+ Geotextile (8.5 m.)	GP	0	NP	34	0	5% Settlement, Continuing Chips Decomposition.

Note: Peak Phi' and c' are from Consolidated-Undrained tests @ 95% of T-99 Density. $1 kN/m^2 = 20.9$ psf, $1 m = 3.28$ ft, NP = Nonplastic

include special circumstances of low risk, substantial cost savings, and under conditions of careful design evaluation, construction control, drainage, and monitoring. Use of clay-rich backfill material will likely present problems and will perhaps be more trouble and costly than it is worth. However, cohesive soils have good strength properties when kept dry and have been used in some reinforced

structures and reinforced embankments worldwide with moderate success and with significant cost savings (18).

The use of poorly drained backfill materials in reinforced soil structures has been reviewed elsewhere (19). Use of cohesive backfill materials presents design and construction difficulties, making drainage and compaction difficult to achieve, and deformation must be expected and acceptable. The material must be compacted under relatively dry weather conditions and should be compacted dry of optimum. Ideally the soil should be encapsulated to disperse surface runoff and prevent the saturation of the material through micro-fissures filling with water.

Large surface deformations from plastic embankment materials suggest that the reinforcement should be longer than that used for conventional materials. Inextensible reinforcement can help minimize deformations. Pore pressure buildup may occur, which can reduce the frictional resistance of the backfill, so relatively high factors of safety should be used in design. Poorly drained soils can cause significantly accelerated corrosion rates in materials.

Walls with clayey soils for backfill can be designed, but long-term creep, deformation, and lateral earth pressures are difficult to predict. Walls should be constructed with a batter or a stepped flexible face to accommodate the expected deformation. Expansive clays should be avoided or modified. Forces on face connections may be relatively large. Clay-rich material can more successfully be used in reinforced fills than walls because face deformation is seldom an issue in fills.

In structures with clay-rich backfill material, the use of frequently spaced reinforcing layers and thick needle-punch nonwoven geotextiles appears desirable to add a "wicking" effect to the structure and allow for some pore pressure dissipation, especially during construction.

SELECTED CASE HISTORIES

Case History 1

Site: Stump Springs #11, #14, Sierra National Forest, Shaver Lake, California: Date constructed: 1983, Wall type: welded wire, 1:6 batter, Wall height: 4.5–6.8 m (15–22 ft).

The backfill material is fine silty sand (SM) to a clayey sand (SC) with up to 42 percent fines and a PI of 15. The soil is a fine decomposed granite, and specified compaction was 95 percent of AASHTO T-99.

Numerous slides, washouts, and roadway fill failures occurred during a major storm. Repairs included construction of a reinforced fill, two concrete crib walls, and five welded wire walls. Welded wire walls were chosen because of their relatively low cost, flexibility, and ease of construction. Manufacturers' designs were selected, and geocomposite drains were installed behind the walls and flow was monitored.

To date all walls and drainage systems have performed well. No sign of deformation is seen on the paved road above the welded wire walls. Overall settlement is 3 percent of the wall height, resulting in some typical face bulging in the wire in specific lifts.

Case History 2

Site: Grave site, Plumas National Forest, Oroville, California: Date constructed: 1987, Wall type: geotextile wall 1:6 batter, Wall height: 2.7 m (9 ft).

The backfill material is coarse to fine, nonplastic silty sand (SM) with up to 26 percent fines. Soil is a decomposed granite excavated on site, compacted to 90 percent of T-180. A geocomposite drain was placed behind the backfill.

The site was a small roadway fill failure that occurred as a result of a heavy rain. A retaining structure was needed to provide adequate roadway width and keep the toe of the fill out of a creek. A geotextile wall was chosen because of the low cost, remote area, and

simplicity of construction. A lightweight needle-punch, nonwoven geotextile was used, placed in 15- and 23-cm (6- and 9-in.) lifts, and treated with asphalt emulsion on the face. The Forest Service design procedure was used.

To date the wall has performed well with some bulging of each lift, face irregularity, and overall 7 percent face settlement. Exposed face geotextile has deteriorated slightly, and animals have chewed small holes in the face, but with no loss of backfill material.

Case History 3

Site: Butt Valley, Plumas National Forest, Canyon Dam, California: Date constructed: 1988, Wall type: tire-faced geotextile wall, Wall height: 3.1 m (10 ft).

The backfill material is a local gravelly clayey sand, with 30 to 38 percent fines and a PI of 8 to 9, derived from metamorphic rock. Field compaction was 93 to 95 percent of AASHTO T-99.

This small roadway fill failure area required a retaining structure. Since a local contractor had access to used tires and several "gravity" tire structures had been recently built, an MSB wall was designed, using tires as the facing material. A lightweight slit-film woven geotextile was chosen for reinforcement and placed between each two layers of tires. Sixteen vertical rows of tires were used. Tires are held in place by friction between tire layers or the geotextile. The bottom half of the wall is vertical, but because of soil loss around the tires, the upper half has a 1:4 batter.

Face settlement has been surveyed and monitored since construction. No settlement is evident in the roadway on top of the wall, but midwall face settlement has been about 10 percent of the wall height. Most settlement occurred in the first 2 years after construction. Today the wall appears stable and partially vegetated.

Case History 4

Site: Camp 5 Hill, Willamette National Forest, Oakridge, Oregon: Date constructed: 1984, Wall type: geotextile wall with lightweight fill (wood chips), Wall height: 8.5 m (28 ft).

The backfill material is wood chips [7.5 cm max. (3 in.)], having a friction angle of about 34 degrees and dry unit weight of 6.3 kN/m^3 (40 pcf). Procedural compaction was used, with a specified number of roller passes per lift.

The lightweight geotextile reinforced wall was constructed in a large, active slide area where bearing pressure needed to be minimized, the flexibility of a geotextile wall was advantageous, and the wall could conform to the site. The custom design was based on the frictional characteristics of the chips, using a Forest Service–developed design procedure. The slit-film woven geotextile was sprayed with asphalt emulsion for protection, but poor adhesion resulted with this type of geotextile.

To date the geotextile wall has performed well and overall settlement has been about 0.5 m (1.5 ft), or 5 percent of the total height. Gradual settlement continues as the chips decompose with some moisture in the chips. The geotextile is disintegrating, but because of minimal face pressure, no chip loss has occurred. The historic slide has not moved.

Case History 5

Site: Gallatin Marina, Lassen National Forest, Eagle Lake, California: Date constructed: 1989,

Wall type: rigid concrete panel facing with welded wire soil reinforcement, Wall height: 3.8 m (12.5 ft).

The backfill material is a well-graded to poorly graded, nonplastic, coarse sandy gravel (GW) with minimal fines. The materials are soft lightweight volcanic cinders that partially break down during compaction, doubling the percentage of sand-size particles.

The Hilfiker Stabilized Embankment precast concrete panel wall type with a smooth, gravel-textured face, was selected and designed by the contractor to satisfy the need for a durable, aesthetic wall facing along a marina walkway. Changed site conditions caused modification of the initial wall design and materials source. Panels are 0.8 by 3.8 m (2.5 by 12.5 ft), connected to the welded wire reinforcement with pins. The soft foundation was overexcavated to bedrock and backfilled with cinders. A geocomposite drain was installed behind the backfill. Compaction was difficult to control with the lightweight backfill material but was reasonably achieved.

To date the wall looks good and has performed well, with only minor face deformation. With the rigid panels and the lightweight variable backfill material used, the minor face deformation has caused some offset and cracking of a couple panel corners.

Case History 6

Site: Little North Fork, Plumas National Forest, Oroville, California: Date constructed: 1989, Wall type: 1:1 reinforced fill plus welded wire toe wall, Fill height: 15.2 m (50 ft), Wall height: 3.2 m (10.5 ft).

The backfill material is local fine silty sand (SM) to a sandy silt (ML) with up to 55 percent fines and a PI of 3. The soil is derived from weathered metamorphic rock. Specified compaction was 95 percent of AASHTO T-99.

This steep slide area and road repair was investigated and evaluated for a retaining wall but had marginal foundation materials and an unsafe excavation back slope. It was also marginally steep to support a 1:1 fill. Thus a small welded wire retaining wall was custom designed and placed on a firm bedrock area to support the toe of the reinforced fill. A geogrid reinforced fill was designed and constructed above the wall.

To date the composite structure has performed well, with little wall settlement and with good vegetative growth on the fill face except for several local shallow fill face slumps between the layers of secondary reinforcement. The geocomposite drainage system under the fill continues to function, discharging a moderate flow of water.

Case History 7

Site: Thomjack, Klamath National Forest, Yreka, California: Date constructed: 1989, Wall type: timber-faced geogrid wall, Wall height: 4.6 m (15 ft).

The backfill material is a nonplastic silty sand with gravel (SM) and up to 27 percent fines.

A timber-faced structure was chosen to satisfy a natural, aesthetic appearance for this forest road. The wall was custom designed, using a biaxial geogrid for reinforcement and 20- by 20-cm (8- by 8-in.) treated timbers, set on a 1:32 batter. Timbers are connected to the geogrid with staples or the geogrid is sandwiched between the timbers.

To date there has been no visible settlement and overall performance has been excellent. Its rustic appearance is pleasing. No appreciable volume of backfill has been lost through the partially open face.

SUMMARY AND CONCLUSIONS

Hundreds of MSB structures, with a wide variety of construction and facing materials, using local materials have been successfully constructed on Forest Service roads in rural areas over the past 20 years.

Welded wire walls and geotextile or geogrid-reinforced soil walls with various facing materials (such as timbers, gabions, tires, geocells, or segmental concrete blocks) appear ideal for rural applications on low- to moderate-volume roads. Reinforced fills can offer an economical alternative to conventional retaining structures. These structures represent the low range of costs for retaining structures available today and are appropriate in many settings. Substantial cost savings can be realized by their use, not only for the federal government, but also for state transportation agencies, counties, and the private sector.

Use of in-house geotechnical skills is cost effective for projects involving retaining structures. For actual structural design, either a custom design or a standard design provided by vendors may be suitable. However, to evaluate the most applicable type of structure, to perform site evaluation and foundation assessment, to evaluate design and construction modifications, and to perform needed external and global stability analysis (typically not provided by vendors), timely input from qualified geotechnical personnel is necessary.

Finally, significant additional cost savings can be realized by using local, typically on-site backfill material. Its use by the Forest Service has been satisfactory. However, use of marginal materials introduces the need for positive drainage, some additional construction effort, and allowance for some settlement or overall deformation. In most noncritical applications, these factors are acceptable and economical. Use of marginal materials will likely become more widely accepted.

ACKNOWLEDGMENTS

The author thanks Jim McKean, Ken Inouye, Ed Rose, Michael Burke, Mark Truebe, Richard Vandyke, and John Mohney, Forest Service geotechnical engineers, for their assistance in providing technical information and suggestions for this paper.

REFERENCES

1. Mohney, J. *Retaining Wall Design Guide—Forest Service.* FHWA-FLP-94-006, U.S. Department of Agriculture, Updated 1994, 347p.
2. Steward, J., R. Williamson, and J. Mohney. *Guidelines for Use of Fabrics in Construction and Maintenance of Low-Volume Roads.* FHWA-TS-78-205. USDA Forest Service, 1977, 153pp.
3. Mitchell, J. K., and W. Villet. *NCHRP Report 290 Reinforcement of Earth Slopes and Embankments.* TRB, National Research Council, Washington, D.C., 1987, 323p.
4. Prellwitz, R., T. Kohler, and J. Steward. *Slope Stability Reference Guide for National Forests in the United States.* EM-7170-13, Forest Service, U.S. Department of Agriculture, 1994, 1,091pp.
5. Christopher, B. R., S. Gill, J. Giroud, I. Juran, J. K. Mitchell, F. Schlosser, and J. Dunnicliff. *Reinforced Soil Structures, Volume I: Design and Construction Guidelines.* FHWA-RD-89-043. FHWA, U.S. Department of Transportation, 1990, 301pp.

6. Wu, J. T. H. *Design and Construction of Low Cost Retaining Walls—The Next Generation in Technology*. CTI-UCD-1-94, Colorado Transportation Institute, Denver, 1994, 152p.
7. Stuart, E., K. Inouye, and J. McKean. Field and Laboratory Evaluation of Geocomposite Drain Systems for Use on Low-Volume Roads. In *Transportation Research Record 1291*, Vol. 2. TRB, NRC, Washington, D.C., 1991, pp. 159–165.
8. Koerner, R. M. Designing for Flow. *Civil Engineering*, Oct. 1986, pp. 60–62.
9. Barrett, R. K. Geotextiles in Earth Reinforcement. *Geotechnical Fabrics Report*, March–April 1985, pp. 15–19.
10. Christopher, B. R., C. Bonczkiewicz, and R. D. Holtz. Design, Construction, and Monitoring of Full Scale Test of Reinforced Soil Walls and Slopes. *Proc. Recent Case Histories of Permanent Geosynthetic-Reinforced Soil Retaining Walls*. A. A. Balkema, Rotterdam, the Netherlands, 1994, pp. 45–60.
11. Allen, T. Issues Regarding Design and Specification of Segmental Block-Faced Geosynthetic Walls. In *Transportation Research Record 1414*, TRB, National Research Council, Washington, D.C., 1993, pp. 6–11.
12. Keller, G., and O. Cummins. Tire Retaining Structures. *Engineering Field Notes*, Vol. 22, ETIS, U.S. Department Agriculture Forest Service, March–April 1990, pp. 15–24.
13. Bathurst, R. J., and R. E. Crowe. Recent Case Histories of Flexible Geocell Retaining Walls in North America. Tatsuoka and Leshchinsky, eds. *Proc., Recent Case Histories of Permanent Geosynthetic-Reinforced Soil Retaining Walls*. A. A. Balkema, Rotterdam, the Netherlands, 1994, pp. 3–19.
14. Design Guidelines for Use of Extensible Reinforcements (Geosynthetics) for Mechanically Stabilized Earth Walls in Permanent Applications. *Task Force 27 Report: In Situ Soil Improvement Techniques*. AASHTO, Washington, D.C., Aug., 1990, 324 p.
15. Segrestin, P. GRS Structures with Short Reinforcements and Rigid Facing-Discussion. Tatsuoka and Leshchinsky eds. *Proc., Recent Case Histories of Permanent Geosynthetic-Reinforced Soil Retaining Walls*. A. A. Balkema, Rotterdam, the Netherlands, 1994, pp. 317–322.
16. Huang, C. C. Report on Three Unsuccessful Reinforced Walls. Tatsuoka and Leshchinsky, eds. *Proc., Recent Case Histories of Permanent Geosynthetic-Reinforced Soil Retaining Walls*. A. A. Balkema, Rotterdam, the Netherlands, 1994, pp. 219–222.
17. Wu, J. T. H. Measured Behavior of the Denver Walls. *Proc., International Symposium on Geosynthetic-Reinforced Retaining Walls*. A. A. Balkema, Rotterdam, the Netherlands, 1992, pp. 31–42.
18. Berdago, D. T., R. Shivashankar, C. L. Sampaco, M. C. Alfaro, and L. R. Anderson. Behavior of a Welded Wire Wall with Poor Quality Cohesive, Friction Backfills on Soft Bangkok Clay: A Case Study. *Canadian Geotechnical Journal*, Vol. 28, 1991, pp. 860–880.
19. Zornberg, J., and J. K. Mitchell. *Poorly Draining Backfills for Reinforced Soil Structures—A State of the Art Review*. Geotechnical Engineering Report UCB/GT/92-10, Department of Civil Engineering, University of California, Berkeley, 1992, 101p.

Publication of this paper sponsored by Committee on Geosynthetics.

Retaining Structure Selection at Project Level

TIMOTHY G. HESS AND TERESA M. ADAMS

With recent advances in retaining structure technology, the retaining structure selection process has become increasingly complicated. Because of the potential for cost savings, the benefits of optimizing the selection process are significant. Eight state departments of transportation (DOTs) are characterized in terms of how they select and analyze designs of, and obtain subject-matter expertise on, retaining structures. Results indicate most retaining structure types are specified by DOT engineers, including those designed by engineering consultants. However, 90 percent of the DOTs select only retaining structures for which they have in-house design expertise. The rational is presented for DOTs to have design expertise within their agencies for the full range of retaining structure technologies available if they are to select optimal structure types.

Recent advances in retaining structure technology complicate the selection of retaining structures at the project level. In the past, retaining structure options consisted of a limited number of externally stabilized structures with few other choices available. Today, the decision maker has numerous options encompassing internally and externally stabilized structures. As a result, retaining structure selection has become an optimization problem covering not only new technologies but also new engineering concepts. To select the optimum structure for a given project, the decision maker must have expertise in a wide variety of retaining structure technologies or access to subject-matter expertise.

The benefits of optimizing the retaining structure selection process are tremendous. State departments of transportation (DOTs) collectively overspend approximately $700 million per year by not optimizing decisions to take advantage of new retaining structure technologies and materials (1).

As part of an effort to characterize the retaining structure selection process, questionnaires were sent to subject-matter experts at eight state DOTs: California, Colorado, Florida, Kentucky, New York, Ohio, Texas, and Washington. Six of the experts are geotechnical engineers and two are structural engineers. These DOTs are considered national leaders in the application of diverse retaining structure systems (J. DiMaggio, FHWA, personal communication with T. M. Adams, July 1993). The data collected are representative across the United States. All the questionnaires were completed. Subsequently, six of the experts explained their responses during telephone interviews.

Varying design and construction practices influence the selection process. Each expert estimated the percentage of retaining structures designed by DOT engineers and consultants. Table 1 summarizes the

T. G. Hess, Department of Civil and Environmental Engineering, University of Wisconsin-Madison, Current affiliation: U.S. Army Corps of Engineers, Rock Island District, Design Branch, P.O. Box 2004, Rock Island, Ill. 61204. T. M. Adams, Department of Civil and Environmental Engineering, University of Wisconsin-Madison, 2208 Engineering Bldg., 1415 Johnson Dr., Madison, Wis. 53706.

results. Three of the eight states use DOT engineers to complete the design and construction plans for less than 25 percent of their retaining structure projects. All three of these states are east of the Mississippi River. In the other five states, DOT engineers complete the design and construction plans for 50 to 100 percent of the projects. All but one of these states are west of the Mississippi River.

The responses to the questionnaire and telephone surveys were analyzed. The results indicate that it is essential for state DOTs to develop and maintain subject-matter experts for the full range of retaining structure solutions and to build communication channels between subject-matter experts and project engineers.

OBJECTIVE

This paper characterizes the selection of earth-retaining structures associated with highway projects. In particular, the paper focuses on characteristics of the decision maker, the decision process, the information available at the time of the decision, the importance of project parameters, and the impact of available expertise at the time of the decision. The objective is to answer the following questions about retaining wall selection at the project level.

- Who selects the type of earth-retaining structure to construct?
- Are selection decisions made in house (within the DOT) or by consultants?
- Does the decision maker's lack of expertise restrict the outcome of the decision?
- At what stages of highway construction are retaining walls selected?
- Is a formal decision process used to select retaining walls?
- Who designs the wall and prepares the construction plans?
- Who prepares preliminary and detailed cost estimates for retaining structures?
- Who determines the construction methods and specifies the construction materials?
- Are retaining structure decisions optimized by value engineering?
- Do decision makers receive comments on constructability from the field?
- What project parameters influence the type of wall selected?
- What information is available at the time of retaining structure selection?

The availability of new economical retaining structure systems stimulates the desire to optimize the choice of retaining structure. Obviously, the decision maker has control. Some questions gathered information on whether a geotechnical engineer, structural engineer, or project manager is the decision maker. Other questions determined whether professional consultants are included in the decision process and under what circumstances.

TABLE 1 Design Practices by State

States	Percent of Retaining Structures Designed by DOT Engineers
FL, KY, OH	0-25%
	26-50%
CO, NY TX	51-75%
CA, WA	76-100%

For simplicity, a consistent terminology was adopted. In this paper, *project manager* refers to engineers who are project managers or project engineers. *Design* includes both engineering design and engineering analysis. The scope of retaining structures is earth-retaining walls; thus in this paper, *structure* and *wall* are synonyms.

SCOPE OF RETAINING STRUCTURES

In spring 1993, a survey for collecting knowledge about selecting earth-retaining structures for highway projects was developed. The survey covered 24 conceptual retaining structures included in a formalized retaining wall selection procedure described in the Colorado DOT bridge design manual (2). Twelve retaining structure experts, including owners, consultants, contractors, and educators, completed the survey. Considering their suggestions, six wall designs were deleted and five wall designs were added. The revised scope of retaining structures was presented to 12 additional experts in January 1994. Considering their responses, the scope of walls was edited again. Five more walls were deleted from the list. Table 2 contains the final set of 18 retaining structures.

Eight state DOTs ranked the 18 walls in Table 2 according to frequency of use, with 1 being most frequent. A ranking of 18 indicates most infrequently or never used. Table 2 lists the 18 retaining

TABLE 2 Ranking of Retaining Structures From Most Frequently Used to Least Frequently Used

Description	Stabilization Method	Average Ranking
Mechanically stabilized earth wall. Select fill reinforced earth with strips, mats, or grids of metal or geosynthetic tensile reinforcements	I	1.9
Shallow embedded cantilever wall with tiebacks anchored to the stabilized zone	E	6.6
Cast-in-place cantilever T-wall	E	7.4
Soldier piles. Cantilevered H-piles (driven or placed in drilled caissons) with wood or precast concrete lagging	E	8.5
Shallow embedded cantilever wall with deadman anchors	E	9.6
Gabions. Welded wire baskets filed with course aggregate stone	E	10.3
Cast-in-place L-wall or invert L-wall	E	10.6
Crib wall. Single or double, step-front or step-back crib wall constructed of precast concrete or lumber stringers and tie members	E	11.5
Metallic bin walls. Corrugated aluminum or steel bins and in-filled soil form composite material	E	11.6
Mass cast-in-place concrete gravity wall	E	11.9
Modular wall. Precast/prefabricated. Most are proprietary. Modular units and in-filled soil form composite material	I	11.9
Sheet pile. Embedded sheet pile cantilever wall	E	13.5
Soil nailed wall. Facing covered cuts with uniformly spaced top-to-bottom drilled or driven nails	I	13.8
Cantilever T-wall with precast post-tensioned stem	E	14.4
Drilled caissons. Embedded cantilever wall constructed of contiguous, secant or tangent drilled caissons back-filled with concrete. With or without lagging	E	15.0
Multi-anchored facing wall. Precast concrete multi-anchored facings with tiebacks anchored to the stabilized zone or fill	E	15.3
Multi-anchored facing wall. Creeping slopes doweled with caissons or piles for stability. Precast concrete facings are anchored to the dowels	E	15.5
Diaphragm wall. Embedded cantilever wall constructed of a trenched slurry concrete diaphragm wall	E	16.8

E=externally stabilized structure, I=internally stabilized structure

structure types ordered from most to least frequently used based on average ranking.

The wall types in Table 2 are internally or externally stabilized. Internally stabilized walls rely on the soil itself for stability and are often considered geotechnical solutions. Externally stabilized walls use structural mechanisms for stability and are usually regarded as structural solutions.

The most frequently used retaining structure is the mechanically stabilized earth wall. Various externally stabilized cantilever structures comprise the remaining four most frequently used wall types. Three infrequently used wall types are diaphragm walls and doweled or tieback multianchored facing walls.

CHARACTERISTICS OF DECISION MAKER

Each DOT identified the project participants (DOT engineer, consultant, or contractor) who select the wall type, complete the design, determine construction methods, specify materials, and prepare cost estimates for retaining structures associated with highway projects. The results characterize the overall responsibility of each participant for retaining structure selection and design.

Table 3 shows the percentage of responses indicating the responsible project participant for various activities. At some DOTs, more than one participant performs an activity. For example, the type of retaining structure selected by a DOT engineer or consultant may depend on whether the wall is also designed by the DOT engineer or the consultant. In these cases, the responses were distributed so that the results are evenly weighted among the eight DOTs. Three-fourths of the responses indicate that a DOT engineer selects the type of retaining structure. Only half the responses indicate that design, analysis, and construction plans are completed by a DOT engineer. This means many retaining structures are selected by DOT engineers, then designed by consultants or contractors. Data collected during the interviews explain this. Six experts say their DOT never seeks the services of a consultant solely for selecting the type of retaining structure. Consultants select and design or design only. The data presented in Table 3 include all retaining structure projects, regardless of whether DOT engineers or consultants prepare the design and construction plans.

Table 4 shows a breakdown of who selects the wall type for projects sent to a consultant for design. Half the responses indicate a DOT engineer decides the type of retaining structure for these pro-

jects. The most frequent decision maker within the DOT is the project manager. Project managers are responsible for the project schedule and project coordination. Project managers, usually in the DOT roadway design division at the regional level, often have backgrounds in geometric design. Otherwise a project manager's background may be in a variety of other engineering disciplines. This suggests a need for direct communication between project managers and subject-matter experts for selecting wall types.

According to Tables 3 and 4, 12 percent of the responses indicate the construction contractor selects the retaining structure. This percentage is probably not representative of all state DOTs, but because of the unique contracting practices of one of the DOTs interviewed. This particular DOT requires the construction contractor to select and design all retaining structures associated with the project. The contractor must select a retaining structure from a preapproved list and hires a consultant to complete the design.

Each DOT indicated by discipline who within the DOT performs planning and design activities when these activities are done by DOT engineers. Table 5 shows the results. At some DOTs, various participants perform the activities depending on the scope of the project. Thus, some DOTs indicated more than one discipline. In these cases, responses were adjusted accordingly. As shown in Table 5, half the responses indicate that the project manager selects the type of retaining structure. If the schedule permits, the project manager usually makes the decision on the basis of advice from a geotechnical or structural team member and sometimes from both. About 40 percent of the responses indicate that the structural engineer selects the type of retaining structure, and less than 10 percent indicate the geotechnical engineer is directly responsible for selecting the type of retaining structure. Typically at a DOT, most project managers are within the roadway design division and most structural engineers are within the bridge division. Because these engineers are selecting the type of retaining structure, it is imperative that they have knowledge in the full range of retaining structure technologies available or involve subject-matter experts from other divisions.

When asked who completes the design and prepares construction plans for retaining walls, a third of the responses indicate a geotechnical engineer, and twice as many indicate a structural engineer. When this information is considered along with the frequency of wall type in Table 2, it appears structural engineers within the bridge divisions are not strictly selecting and designing traditional externally stabilized retaining structures. At several DOTs studied,

TABLE 3 Planning and Design Activities by Project Participant

Project Activity	Percent of Responses*		
	DOT Engineer	Consultant	Contractor
Select Type of Retaining Structure	76	12	12
Complete Design and Prepare Construction Plans	54	46	0
Determine Major Construction Methods	43	7	50
Specify Construction Materials	92	8	0
Prepare Preliminary Cost Estimate	72	28	0
Prepare Detailed Cost Estimate	88	12	0

*multiple responses are uniformly distributed so that results are evenly weighed among DOTs

TABLE 4   Retaining Structure Selection by Discipline When Design is Contracted Out

Discipline	Percent of Responses*
DOT Geotechnical Engineer	12
DOT Structural Engineer	12
DOT Project Manager/Engineer	26
Consultant	38
Contractor	12

*multiple responses are uniformly distributed so that results are evenly weighed among DOTs

structural engineers in the bridge divisions design all retaining structures, externally and internally stabilized. At other DOTs, however, internally stabilized structures are designed solely by geotechnical engineers in the geotechnical divisions, with the bridge division having responsibility for external solutions involving steel and concrete. Less than 10 percent of the responses indicate the project manager completes the design and plans. A project manager completes the design and construction plans for sites conducive to standard plans.

The division of the organization is not particularly important as long as optimal solutions are designed. It is important for retaining structure types to be selected and designed by an engineer or team of engineers with knowledge of the full range of retaining structure technologies available. Otherwise a DOT cannot be sure it is specifying optimal retaining structure solutions.

CHARACTERISTICS OF DECISION PROCESS

Characteristics of the decision process include the project phase during which the retaining structure type is selected, the selection procedure, and the impacts of value engineering and constructability comments on the decision process. Each DOT surveyed provided information about these characteristics.

Each DOT indicated the phase of the project during which the type of retaining structure is selected. Responses indicating multiple phases were uniformly distributed so that results are evenly weighed among the DOTs. As shown in Table 6, almost three-fourths of the responses indicate that the wall type is selected during preliminary or final design. Few responses indicate that the structure type is selected during the planning phase. About one-fourth of the responses indicate that the retaining structure type is selected during the construction phase.

There are three scenarios for selecting a retaining structure during the construction phase. The first involves the use of a proprietary retaining wall. The use of a proprietary wall is usually determined during the design phase. The DOT specifies on the contract drawings that the contractor will choose a wall system from a list of previously approved proprietary walls. Typically, the list includes at least three different retaining wall systems. The second scenario involves value engineering. For six of the eight states surveyed, value engineering studies may change the type of retaining structure constructed. For 50 percent of the states, a value engineering study team applies value engineering during design. For the other 50 percent, value engineering studies are conducted during construction by the construction contractor. If a contractor proposes a significantly less expensive structure that is equal to the specified structure, the proposal is usually accepted and the type of retaining structure is changed. The third scenario was described earlier. One of the DOTs requires the construction contractor to select the type of retaining structure from a preapproved list.

The states indicated whether the types of retaining structures selected are never, rarely, frequently, or always on the basis of a formal or informal decision process. A formal decision process includes a formal design report describing several retaining structure alternatives and recommendations based upon some decision analysis. An informal analysis is not documented in a formal report. As shown in Table 7, the responses were distributed. Two states frequently prepare a formal report, and an informal analysis is always or frequently done. One state rarely completes a formal report or informal analysis because the contractor selects the structure type. The remaining five states prepare a formal design report or an informal analysis, but not both. As a result, half the states frequently or always prepare a formal design report and half do not.

The states were asked how often the engineer, who selects the type of retaining structure, receives comments on the constructability of different retaining structures. Table 8 summarizes the results. Only about one-third of the DOTs have programs that facilitate con-

TABLE 5   Planning and Design Activities at DOTs by Engineering Discipline

Project Activity	Percent of Responses*			
	Geotechnical	Structural	Project	Cost
Select Type of Retaining Structure	8	42	50	0
Complete Design and Prepare Construction Plans	29	63	8	0
Determine Major Construction Methods	24	66	10	0
Specify Construction Materials	50	50	0	0
Prepare Preliminary Cost Estimate	12	39	49	0
Prepare Detailed Cost Estimate	0	14	43	43

*multiple responses are uniformly distributed so that results are evenly weighed among DOTs

TABLE 6 Phase of Project When Retaining Structures are Selected

Phase	Percent of Responses*
Planning	4
Preliminary Design	58
Final Design	15
Construction	23

*multiple responses are uniformly distributed so that results are evenly weighed among DOTs

structability comments from the field. Some of the respondents indicate they receive constructability comments only when there are problems; otherwise constructability comments are rare.

AVAILABILITY OF INFORMATION AT TIME OF RETAINING STRUCTURE SELECTION

Performance of internally stabilized wall systems depends on the in situ soil conditions or properties of available fill. Wall selection decisions for externally stabilized retaining structures depend as well on local site and soil conditions.

Each state indicated whether certain information is always, usually, or never available at the time of retaining structure selection. Table 9 summarizes the responses. The experts who indicate that soil and water table data are never available noted that during selection these parameters may be available for nearby sites.

IMPORTANCE OF PROJECT PARAMETERS

Each expert ranked by priority six project decision parameters on a scale of 1 to 6, where 1 is most important and 6 is least important. Table 10 contains the priority ranking and average rankings. On average, estimated cost and cut versus fill application are the most important decision parameters. All eight DOTs priority ranked one of these parameters as first or second. However, four DOTs also priority ranked one of these parameters very low. On average, expected deflection and aesthetics are least important of the six parameters. For several states, aesthetics is not a limiting factor because it is possible to add an architectural facia to many retaining structures.

IMPACT OF DESIGN EXPERTISE ON RETAINING STRUCTURE SELECTION

The DOTs characterized the impact of design expertise of DOT and consulting engineers on the type of retaining structure selected. The states estimated the impact of designer expertise on the selection process, then the frequency a DOT engineer or consultant has design expertise for the type of retaining structure selected. The results indicate the effect of design expertise on the selection process.

Each DOT was asked whether personal expertise of DOT engineers affects the type of retaining structure selected. The compiled results in Table 11 show almost two-thirds responded that in-house design expertise rarely or never affects the selection of retaining structure type. Ninety percent of responses indicate DOT engineers frequently or always have design expertise for the structure selected. This includes all projects designed in house or by a consultant. Only about 10 percent of responses indicate the DOT rarely has in-house design expertise for the structure type. Thus, most retaining structures being selected for highway projects are those for which DOTs have design expertise.

The DOTs described the role of engineering consultants. Table 12 contains the results. Fifty percent of the respondents are uncertain whether design expertise of their consultants influences the type of wall selected. The other 50 percent believes design expertise of consultants rarely or frequently affects the structure type. Each DOT estimated the frequency that prime consultants have design expertise for structures the consultant selects and designs. Much like the results in Table 11, almost 90 percent of the responses indicate consultants frequently or always have design expertise for the type of retaining structure selected. The states estimated the frequency a prime consultant selects the structure type and then, because of lack of expertise, obtains the services of a subconsultant to complete the design. Prime consultants more frequently subcontract the all-geotechnical or structural work, including selection of the structure types.

OPTIMIZATION OF SELECTION PROCESS

The last two decades brought the introduction and growing use of internally stabilized retaining structures, such as mechanically stabilized earth walls, modular walls, and a variety of ground improvement techniques. From the results presented, DOT engineers select most retaining structures from the types for which they have expertise. For the current paradigm to produce optimal solutions, it is essential for DOTs to have expertise in the full range of retaining structure technologies available. Consequently, highway agencies are challenged to develop and maintain in-house design

TABLE 7 Frequency of Different Retaining Structure Selection Methods

Retaining Wall Selection Method	Percent of Responses			
	Never	Rarely	Frequently	Always
Wall type is selected based on formal report comparing several wall types.	12	38	38	12
Wall type is selected by informal analysis using designer's expertise.	12	25	38	25

TABLE 8 Frequency DOT Engineers Receive Constructability Feedback from Field

Frequency	Percent of Responses
don't know	0
never	0
rarely	38
sometimes	24
frequently	38
always	0

expertise for all feasible retaining structure solutions (3). Because of the tremendous cost savings, the state DOT must develop and rigorously maintain its knowledge base.

Alternatively, DOTs could move the retaining structure selection decision out of the DOT. Increasing reliance on design consultants and proprietary retaining wall vendors accomplishes this. However, there are two problems with this approach. The first occurs because consultants rarely or never select retaining structures for which they have no design expertise. To optimize the selection process, it would be necessary for state DOTs to require engineering consultants to have design expertise in most or all of the feasible retaining structure solutions. Although this might not be a problem for some larger consultants, many otherwise technically capable consultants would be restricted from state DOT design contracts. This would severely restrict the state's choices in selecting consultants for retaining structure design and would not be in the best interest of the highway agency. Second, proprietary retaining wall vendors promote and build one or a limited selection of retaining wall systems. Now, many DOTs have excellent success using proprietary wall systems after DOT engineers identify the appropriate proprietary system. Most proprietary wall systems are based on specific technology not applicable to the entire range of grade separation problems. A proprietor should not be expected to understand all alternative technologies.

An alternative to acquiring and maintaining in-house expertise is for state DOTs to rely on competitive bidding. At one of the DOTs interviewed, most retaining structures are selected by construction contractors. The DOT engineers indicate the lines, grades, and location of a retaining structure on the construction plans without indi-

cating the type of structure. As part of the bid proposal, the contractor selects the most cost-effective retaining structure from a group of preapproved wall systems. This particular DOT has a large number of preapproved retaining structures, so the competitive bidding process usually results in a good choice. The competitive bidding process does not guarantee the best solution. To be competitive, the construction contractor no doubt picks the most cost-effective structure. However, the contractor has no incentive to pick the optimum structure based on other parameters such as durability, maintenance, and least life-cycle cost, unless specifically required by the DOT. To analyze and specify parameters other than cost, DOTs need in-house expertise.

The best solution for optimizing retaining structure selection is for a state DOT to select the type of retaining structure with in-house DOT engineers. Once the type of structure is selected, the state can design the structure and prepare construction plans with DOT engineers or consultants as appropriate. One exception is a highway project entirely planned and designed by a consultant. In this case, the retaining structure can be selected by the consultant if the consultant has the necessary expertise.

CONCLUSION

As part of an effort to characterize the selection process for earth-retaining structures on highway projects, information was collected from subject-matter experts at eight state DOTs. This information focused on the characteristics of the decision maker, the decision process, the availability of information at the time of decision, the importance of project parameters, and the impact of design expertise. The results reveal some interesting patterns. First, consultant services are never used exclusively to select the type of retaining structure. Half the responses indicate consultants or design-build contractors design and prepare construction plans for retaining structure types specified by DOT engineers. DOTs and their consultants always or frequently have the design expertise for the type of retaining structure selected.

Recent advances in retaining structure technology bring the introduction and use of numerous new retaining structure systems. These advances require state highway agencies to develop and maintain in-house knowledge of new technologies if they are to continue to specify the best retaining structure solutions. State DOTs should assess their knowledge level and ensure they possess up-to-date expertise for current retaining structure technologies. A DOT should develop and maintain expertise within its organization for the full range of retaining structure solutions available today, especially those listed in Table 2.

TABLE 9 Availability of Information at Time of Wall Selection

Project Data	States Surveyed							
	1	2	3	4	5	6	7	8
Soil borings with strata identified	A	N	U	U	U	A	A	A
Water table location	A	N	U	U	U	A	A	A
Soil lab test report	A	N	U	U	U	A	A	A
Horizontal and vertical alignments	A	U	A	A	U	A	A	A

A=always, U=usually, N=never

TABLE 10   Importance of Project Parameters for Selecting Retaining Structures

Project Parameter	States Surveyed								Average
	A	B	C	D	E	F	G	H	
Cut/Fill Application	1	2	6	4	1	2	1	2	2.4
Estimated Cost	5	1	1	2	2	5	5	1	2.8
Tolerance to Settlement	4	3	2	3	5	4	2	4	3.4
Wall Height	2	5	4	5	3	1	4	3	3.5
Expected Deflection	3	6	3	6	6	3	3	5	4.4
Aesthetics	6	4	5	1	4	6	6	6	4.6

TABLE 11   Percentage of Responses Indicating How Design Expertise of DOT Engineers Influences Type of Retaining Structure Selected

Design Expertise	Never	Rarely	Frequently	Always
Expertise of DOT engineers influences type of structure selected	24	38	38	0
DOT engineers have expertise for the type of retaining structure selected	0	12	38	50
DOT engineers do not have expertise for type of retaining structure selected	38	50	12	0

TABLE 12   Percentage of Responses Indicating How Design Expertise of Consultants Influences Type of Retaining Structure Selected

Design Expertise	Don't Know	Never	Rarely	Frequently	Always
Prime consultant's expertise influences type of retaining structure selected	50	0	25	25	0
Prime consultant has expertise for the type of retaining structure selected	0	0	12	63	25
Prime consultant does not have expertise for the type of retaining structure selected and subcontracts design to specialty consultant	12	12	76	0	0

ACKNOWLEDGMENTS

This work was supported by the Colorado DOT Division of Information Services Contract #94437; the NSF Award No. MSS-9301284, Mehmet Tumay, cognizant program director; the U.S. Army Engineers and Scientists Professional Development Program; and the U.S. Army Corps of Engineers, Rock Island District. The authors gratefully acknowledge this support and the assistance of the following: T. Dickson, New York State DOT; J. Dimaggio, FHWA; G. Garofalo, California DOT; D. Greer, Kentucky Transportation Cabinet; D. Jenkins, Washington State DOT; G. Odem, Texas DOT; P. Passe, Florida DOT; M. Riaz, Ohio DOT; J. Siccardi and T. Wang, Colorado DOT; and J. Witham, D'Appolonia Consultants.

REFERENCES

1. Barrett, R. K. Can You Build a Retaining Wall for Less Cost? *Geotechnical Fabrics Report*, Vol. 10, No. 2, March 1992, pp. 14–17.
2. Wall Selection Factors and Procedures (Subsection 5.4) and Work Sheets on Earth Retaining Wall Type Selection (Subsection 5.5). In *Bridge Design Manual*, Staff Bridge Branch, Colorado Department of Transportation, Denver, 1991.
3. Adams, T. M., R. K. Barrett, and T. Wang. Colorado's Knowledge System for Retaining Wall Selection. In *Transportation Research Record 1406*, TRB, National Research Council, Washington D.C., 1993, pp. 19–30.

Publication of this paper sponsored by Committee on Geosynthetics.

Contracting for Mechanically Stabilized Backfill Walls

GEORGE A. MUNFAKH

The key differences in the mechanically stabilized backfill (MSB) walls available in the U.S. market are discussed, particularly with respect to their impact on the walls' stability and long-term performance. These differences are mainly in the soil-reinforcement interaction, the strength and stiffness of the reinforcement, the bond between soil and reinforcement, and the durability of the system. Other factors that make a difference include the strain compatibility between soil and reinforcement, the deformation characteristics of the backfill, and the aesthetic and environmental impacts of the facing. Contracting procedures for MSB wall projects are discussed, and lessons learned from case applications are highlighted with particular reference to contracting methods and economical benefits. A preferred method that would ensure low cost, speedy process, and minimum confrontation between the design engineer and the vendor is recommended.

Mechanically stabilized backfill (MSB) walls and embankments have many advantages over conventional systems. Low cost, simple and rapid construction, no required formwork, construction at low temperatures, aesthetically pleasing facings, flexibility, and tolerance to vertical and horizontal movements make the use of these systems attractive. In recent years, construction of MSB walls and embankments has resulted in substantial savings in cost and right-of-way and marked reductions in environmental impacts, such as when used for embankments adjacent to or crossing wetlands.

The MSB walls come in a wide variety of looks, shapes, sizes, and materials, each promoted by a specialty contractor, a product manufacturer, or a combination of both. Although their basic principle is the same, distinct differences in these systems are serious enough to affect their performance if they are not attended to. These differences also make contracting for such systems a difficult, sometimes frustrating task facing the wall owners and their engineering representatives.

The purpose of this paper is to discuss contracting for MSB walls in the United States. Different procurement procedures will be discussed and lessons learned from several case applications will be presented. The differences in the available wall systems that fall under the MSB category will be discussed and guidelines will be established for use by the design engineer in approving or rejecting a particular wall system proposed by the contractor.

MSB WALLS—AN OVERVIEW

MSB walls are mechanically stabilized earth walls that involve the use of backfill. They are formed basically by the inclusion of reinforcing elements within a compacted backfill behind a vertical or near-vertical wall face. The backfill soil and the reinforcing elements act in unity to form a composite structure that resists the applied loads.

The development of these walls began when Henry Vidal introduced and patented the system of terre armee (reinforced earth) in 1966. The first application of the system was a highway project in Nice, France. Its first U.S. application was a 55-ft-high wall in the San Gabriel Mountains of Southern California. Since then, numerous types of walls have been developed and successfully applied in construction of highways, bridges, railroads, dams, seawalls, and other structures. Table 1 lists most of the MSB walls used in the United States.

An MSB wall has three main components: reinforcing inclusions, backfill, and facing. Different materials (metals, polymers, geotextiles) and shapes (strips, grids, sheets, rods, fibers) have been used for reinforcement. The backfill material usually consists of cohesionless free-draining soil, but other soils have been used with some systems. At the edge of the reinforced backfill, a facing is provided to retain the soil at the face and protect the exposed reinforcing elements from weathering effects. The facings currently used include precast concrete elements, metal sheets and plates, welded wire mesh, concrete blocks, timber, rubber tires, shotcrete, and others.

Construction of MSB walls involves placement of alternating layers of compacted backfill and reinforcement, with each reinforcing element connected to a facing unit or wrapped around the backfill layer at the face. Drains are installed, if needed, and the exposed reinforcement is protected from weathering effects. Before placement of the first backfill layer, the site is prepared and unsuitable soils are removed. Although the general construction approach is the same, certain construction details may differ from one system to another as a result of differences in the reinforcing elements, the wall facings, the labor and equipment requirements, and the experiences of the specialty contractors.

The design of an MSB wall involves determining external and internal stability. For external stability, the backfill and reinforcing elements are considered a coherent body subjected to loads from the in situ soil behind it and any surface loads from traffic, adjacent structures, and so forth. The reinforced-soil block is then analyzed against sliding, overturning, bearing capacity failure, and deep-seated shear failure. The internal stability of the system involves analyzing the tension in the reinforcement, the pullout resistance in the soil-reinforcement interface, and the durability of the reinforcing elements against long-term weathering effects. In seismically active areas, the seismic capacity of the reinforced-soil system is analyzed. The design of MSB walls has been documented in a number of comprehensive references (1–3). Recommended safety factors against internal and external stability considerations are summarized elsewhere (4).

Parsons Brinckerhoff Quade & Douglas, Inc., One Penn Plaza, New York, N.Y. 10119.

TABLE 1   MSB Walls Used in the United States

Wall System	Reinforcement	Facing
Reinforced Earth	Steel Strips	Concrete Panels
VSL Retained Earth	Steel Grid	Concrete Panels
Websol Reinforced Soil System	Plastic Strips	T-Shaped Concrete Panels
Welded Wire Wall	Welded Wire Mesh	Wrapped Around Wire Mesh
Reinforced Soil Embankment	Welded Wire Grid	Conrcete Panels
Eureka Reinforced Soil	Welded Wire Mesh	Cast-in-Place Concrete
Hilfiker Stabilized Embankment	Welded Wire Mesh	Large Smooth Concrete Panels
Tensar Geogrid System	Geosnythetic Grid	Wrapped Grid, Shotcrete, Blocks
Matrix Geogrid Wall	Geogrid Mats	Wire Mesh and Geotextile
USFS Geotextile Wall	Geotextile Sheet	Wrapped Sheets, Shotcrete
CTI Wall	Geosynthetic Grid	Timber
Modular Block Geotextile Wall	Geotextile Sheets	Stacked Concrete Blocks
Mechanically Stabilized Embankment	Steel Bar Mats	Precast Concrete Units
Georgia Stabilized Embankment	Steel Bar Mats	Concrete Panels
Miragrid System	Geosynthetic Grid	Precast Concrete Units
Geocell Wall	Geosynthetic Grid	Cellular Confinement System
Pyramid Modular Block System	Steel Strips, Geostraps	Concrete Blocks
Maccaferri Terramesh System	Steel Wire Mesh Sheets	Rock Filled Wire Baskets
Anchored Earth Wall	Steel Rods	Concrete Panels
Tire-Faced Wall	Geotextile Sheets	Stacked Tires

KEY DIFFERENCES IN MSB WALLS

Although all MSB walls follow the same basic principle and design philosophy, there are distinct differences among the available systems, because of the use of different types and configurations of reinforcing elements, types and geometries of wall facings, and composition and grading of backfill materials. These differences should be carefully evaluated when attempting to substitute one wall system for another.

The key differences in the MSB walls are in the soil-reinforcement interaction and the fundamental aspects of the design, namely the strength and stiffness of the reinforcement, the bond between soil and reinforcement, and the durability and long-term performance of the system. In addition to its impact on the design, changing the wall system may affect other aspects of the project, such as rate of construction, aesthetics, and environmental impact. Following is a brief discussion of the key differences in MSB walls on the market. The differences are mainly the result of changes in the three major components of the reinforced-soil system: the reinforcement, the backfill, and the facing.

Soil-Reinforcement Interaction

The stress transfer between the soil and the reinforcement takes place through one or both of the following interactions: (a) friction along the soil-reinforcement interface and (b) passive soil resistance along the transverse members of the reinforcement.

The relative contribution of each factor depends on the size and configuration of the reinforcement, the soil properties, and the stress-strain characteristics of the system. For strip or sheet reinforcement (reinforced earth, Websol, USFS, CTI, etc.), the interaction between the soil and the reinforcing elements is mainly through friction along the soil-reinforcement interface. In grid-reinforcing systems (Tensar, Welded Wire, VSL, RSE, MSE, GASE, etc.), the pullout resistance is provided by friction and passive soil resistance.

The reinforcing elements are either extensible or inextensible. In inextensible systems (metal or polymer), the strains required to mobilize the full strength of the reinforcing elements are smaller than those needed to mobilize the full strength of the backfill. For extensible materials (geotextile), the required strains are much larger. Therefore, relatively large internal deformations usually occur in these walls. In these cases, the soil's strength properties should be measured at large strains (residual strength). Based on the results of pullout tests, displacements as small as 1.3 mm (0.5 in.) for mobilization of the friction along the reinforcing elements and as large as 100 mm (4 in.) for complete mobilization of the passive soil resistance along the transverse members of the reinforcement are reported (5). Strain compatibility between the soil and the reinforcement is an important factor to be evaluated when comparing two wall systems.

Strength and Stiffness of Reinforcing Elements

The tensile strength of the reinforcement is influenced by its size, shape, arrangement, material characteristics, and a number of external factors, such as temperature, durability, and construction damage. These factors often differ from one system to another.

Where steel is used, the allowable tensile stress is equal to 0.55 F_y (yield stress of steel) for strip reinforcement and 0.48 F_y for grid reinforcement with longitudinal and transverse grid members being of the same size.

In geosynthetic reinforcement, the tensile strength depends on the tensile properties of the load-carrying elements (fibers) and the geometrical arrangement of these elements within the geosynthetic matrix. The tensile characteristics of various load-carrying elements used in geosynthetic materials are illustrated in a work by Lawson (6). With the exception of polyaramid fibers, which exhibit characteristics similar to steel, the stress-strain behaviors of the geosynthetic materials are characterized by lower maximum strengths and higher maximum extensions than those exhibited by steel.

The allowable tension capacity of the geosynthetic reinforcement is influenced by three major factors representing creep, durability, and construction damage. Creep is the increase in extension of a material under a constant applied load that occurs when the material's behavior has reached the plastic state. Because the ambient temperatures of most polymeric-based materials coincide with or are close to their viscoelastic phase, creep becomes a significant factor in assessing their long-term load-carrying capacity. Creep, on the other hand, is not a significant factor when steel reinforcement is used.

Creep reduction factors (defined as the creep limit strength, obtained from creep test results, divided by the ultimate tensile strength) of 0.2 to 0.4 for different types of geotextile are reported (3). At high temperatures, significant creep is experienced by reinforcements made of polyethylene or polypropylene. On the other hand, little change occurs in the load-carrying characteristics of polyestic reinforcement due to temperature.

Placement and compaction of the backfill material against the geosynthetic may reduce its tensile strength. Variations in the installation damage factor of different geosynthetic reinforcements are illustrated elsewhere (6). These variations should be taken into account when an MSB wall with one type of geosynthetic reinforcement is substituted for another.

Bond Between Soil and Reinforcement

Tensile stresses in the reinforcing elements are transferred to the surrounding soil by forming a bond between the soil and the reinforcement. This bond is formed through friction, passive soil resistance, or a combination of both.

The frictional bond is developed along both sides of the section of the reinforcing element in the resisting zone behind the failure plane. To maintain equilibrium, the frictional bond must resist the maximum tensile load carried by the reinforcing element (pullout resistance).

The apparent coefficient of friction between the soil and the reinforcement is a function of the composition and gradation of the backfill material and the shape and material properties of the reinforcing elements. For instance, the apparent coefficient of friction of ribbed steel strips is twice that of flat tape geotextiles.

Durability and Long-Term Performance

The service life of an MSB wall depends to a great extent on the durability of the reinforcements and to a lesser extent on that of the facing elements. The durability of metallic reinforcements is usually measured by their resistance to corrosion. That of geosynthetics is assessed by the resistance to (a) ultraviolet light exposure, (b) hydrolysis in polyester, and (c) oxidation in polyethylene and polypropylene. These durability factors should be carefully evaluated when comparing two types of MSB walls.

The use of an MSB wall with metallic reinforcements in place of one reinforced with geosynthetics should be carefully evaluated in the presence of highly corrosive environments, such as stray DC currents adjacent to railroad tracks or deicing salts in areas with frequent snowfalls.

The durability of geosynthetic reinforcements is more complicated than that of metallic ones. Geosynthetics are generally made of synthetic polymers manufactured by different processes. Four synthetic polymers are usually used in production: polyamide, polyester, polyethylene, and polypropylene. Although all are subject to degradation by exposure to ultraviolet light, their reactions to other durability effects differ from one to another. For instance, although polyester is susceptible to hydrolysis and loss of strength when in contact with water, the other three materials are not affected. On the other hand, thermal oxidation in the presence of heat and oxygen, which tends to cause a breakdown and cross linking of the molecular chain, is mostly felt by polyethylene and polypropylene.

Other Factors That Make a Difference

The performance of an MSB wall is also influenced by certain factors that may be characteristic of a particular system; thus, the wall may be negatively influenced if another system is used. For instance, although most walls use granular backfill, some promote the use of on site materials. Because the granular soils are well drained, the effective normal stress transfer between the reinforcement and the backfill soil would be immediate as each lift of backfill is placed. For the design loads normally associated with MSB walls, the granular soils behave as elastic materials; thus, no post-construction movements are anticipated. If fine-grained soils are used, their poor drainage characteristics may produce high pore water pressures, which delay the transfer of stresses from the soil to the reinforcing elements, thus producing greater loads against the facing and more deformations during construction. This may require a slower construction schedule or result in a lower safety factor during construction.

The reduced dilatancy and internal drainage of the fine-grained soil also affect the long-term stability and deformations of the system. Outward movements of the wall may be experienced from consolidation of the backfill. Long-term seepage forces and freeze-thaw softening effects may also be experienced if a poorly drained, fine-grained soil is used for backfill.

The facings used have different impacts on the performance of the MSB walls. When discrete elements such as concrete panels are used, they provide flexibility to tolerate differential movements without structural distress. Walls with metal facing elements (metal plates or grids), on the other hand, have the disadvantage of a shorter life because of the potential for corrosion of the metal. When metal wires are used (welded wire or gabions), they also have the disadvantages of an uneven surface, exposed backfill, and susceptibility to vandalism. However, they provide good drainage, flexibility, and ability to vegetate the facing.

Aesthetics and environmental impacts are important factors to consider when substituting wall types. Certain facings (metal plates or grids), for instance, may be more economical but not as attractive as the ones originally selected. To reduce traffic noise in environmentally sensitive areas, walls with open and vegetated facings (gabions, welded wire, etc.) are acoustically superior to those using concrete facings. The open nature of the wall face and the foliage covering in some are effective in absorbing the noise hitting them, compared with other walls where the traffic noise is reflected on hard or smooth continuous surfaces.

CONTRACTING FOR MSB WALLS

The earth-retaining structure is usually a part of a large civil engineering project. In most countries around the world, the contracts for MSB walls are awarded on a design-build basis. The terms of reference specify the requirements of the final product using performance-type specifications; it is left to the contractor to select a wall system, design it, detail it, and, ultimately, build it. In the United States, however, the present contracting policies and procedures for civil engineering projects require the engineer to select and design the structure and to prepare detailed plans and specifications to be followed closely by the contractor in the field. The construction manager and the field inspectors make sure this is done. Technical, practical, economical, and political factors affect the wall selection. These factors are discussed elsewhere and a selection process is recommended (7).

Because of the many systems available on the market and in the interest of economy, alternative designs are usually performed for each project. These designs have been made in one of three ways:

- As a design task performed by the design engineer,
- As the result of a value engineering study performed during design or construction, or
- As an alternative design proposed by the contractor.

Because of the specialized nature of the MSB walls, the vendors are often asked to perform the internal design of the system and the design engineer addresses external stability. The design engineer then prepares a set of construction plans and specifications for bidding purposes. Because procurement of proprietary items is usually not allowed on public-sector projects, the bidding documents usually specify a particular system or "proven equal." The general contractor then shops around for the cheapest MSB wall on the market and proposes it as the "equal." As shown in this paper, however, key differences exist among the many systems that can be categorized as MSB walls. These differences would affect the wall's performance and may even result in failure if they were not attended to. The task of the design engineer then would be to ensure that a proposed alternative is a true equal and to recommend the modifications that should be made to make it so. The recommended changes can be in either the design procedures and parameters or the materials used and construction details.

Another method of procurement that has proven beneficial is one in which the engineer designs more than one system and prepares plans and specifications for alternative designs. The contractor is then asked to bid on one or more of the designed alternatives. In this way, the alternative designs will not be questionable and the procurement process will allow fair and equitable competition among qualified specially contractors.

CASE STUDIES

Following are brief case studies documenting contracting procedures used in procurement of MSB wall projects.

Case 1—North Halawa Valley Access Road, Hawaii

To construct the H-3 Highway tunnel through the Koolau Mountain Ridge of the island of Oahu, access roads with extensive retaining walls were needed on both sides of the mountain. Because the retaining walls were to be constructed in mountainous terrain with difficult accessibility, alternatives requiring heavy machinery were ruled out, and the wall selection concentrated on the three most promising alternatives—a reinforced earth wall, a gabion wall, and a geotextile wall. To minimize construction cost, all three alternatives were designed and the prospective bidders were asked but not required to bid on all three. All walls were required to have a service life of 10 years and to be resistant to the moderately to highly acidic in situ soils.

Figure 1 illustrates typical cross sections of the alternatives. Design of the walls has been discussed elsewhere (8). The average bidding price for the geotextile wall was approximately 32 percent less than that for the reinforced earth wall and 42 percent less than that for the gabion wall. The contract was awarded in 1987 for construction of geotextile walls at a bid price of $175/m² ($16.10/ft²). Because all alternatives were designed ahead of time and detailed in the bidding documents, there were no controversial issues and construction proceeded smoothly and expeditiously.

Case 2—Baltimore Central Light Rail Line, Maryland

The Baltimore Central Light Rail Line is a 43-km-long (27-mi) transit facility linking Baltimore County, Baltimore City, and Ann Arundel County in Maryland. At least nine different retaining wall types have been constructed on this project. The bid documents for each segment included alternative retaining wall types to obtain the lowest cost. In addition, two wall types, a tensar wall and a techwall, were proposed and designed by the contractors as cost-cutting alternatives.

The wall alternatives designed and detailed in the bidding documents included MSB walls, gravity-type walls, cast-in-place concrete walls, and others. The MSB walls included reinforced earth and VSL retained earth. For each MSB wall shown, a conventional alternative was included. In each case, however, the MSB alternative received the lowest bid. Reinforced earth walls were selected in three contracts for low bids of $675 to $795/m² ($62 to $73/ft²). VSL retained earth walls received the low bids of $468 to $479/m² ($43 to $44/ft²) in a fourth contract. The bid price for the tensar wall is not available because it was included in the lump-sum bid of a total construction package.

The internal stability calculations of the MSB systems were performed by the vendors and submitted as shop drawings; review and approval of these submittals went smoothly because they were performed according to criteria established in the contract documents. The tensar wall, however, was proposed by the contractor. Because there were no design criteria in the contract documents for this type of wall, a lengthy review process occurred and several discussions took place between the designer and the vendor regarding design issues, factors of safety, and construction details. The proposed design was finally approved after all the designer's requirements were met. Figure 2 shows construction of a tensar wall with a full-height panel facing.

Case 3—Bronx Parking Facility, New York

High retaining walls were needed to construct a car parking lot adjacent to a school in the Bronx, New York. Several alternatives were analyzed in the design stage and a reinforced earth wall was selected and included in the bidding documents. Because proprietary items were not permitted on that project, the contract specifications allowed substitution of the designed MSB wall with a proven equal. The general contractor proposed a wall alternative using geosynthetics for reinforcement and modular blocks for facing. The con-

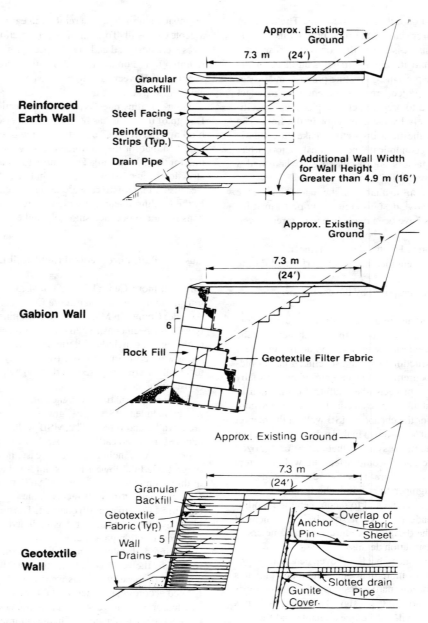

FIGURE 1 Cross sections of wall alternatives.

tractor also proposed on site material for backfill in place of the granular backfill originally specified for the project.

A lengthy review process, including material testing, took place and a number of modifications for the contractor's scheme had to be done before the proposed alternative was accepted (Figure 3). The facts learned from the analyses performed during this evaluation are (a) the backfill should be granular and free draining, (b) uniform compaction is a must, (c) the foundation of the wall facing should be below the frost line and flexible enough to accommodate initial movements, (d) the facing units should have adequate compressive strength and the wall facing should be flexible enough to tolerate vertical and horizontal movements, (e) free drainage immediately behind the wall facing is a must, and (f) the methods of analysis and the safety factors used by the vendors in their designs should be carefully evaluated.

Case 4—Amman–Naur–Dead Sea Highway, Jordan

To cross a landslide area along the Amman–Naur–Dead Sea Highway in Jordan, split-level carriageways were constructed behind retaining walls. Two retaining wall alternatives were considered: a cast-in-place concrete wall and an MSB wall. Because no MSB walls had been built in Jordan before that time and after evaluating previous experiences of the various systems considered, the reinforced earth (RECO) wall was selected by the Jordanian Ministry of Public Works for inclusion in the bidding documents, with no mention of any equal.

After winning the project, however, the general contractor shopped around and proposed an alternative scheme developed by the Hilfiker Corporation, as a modification of their reinforced soil embankment system, to resemble the reinforced earth features

FIGURE 2   Construction of a full-height-panel tensar wall.

FIGURE 4   MSB wall along the Amman–Naur–Dead Sea highway.

FIGURE 3   Geogrid reinforced wall with modular block facing.

included in the bidding documents. No reduction in the bidding price was offered for the proposed alternative.

The internal stability calculations submitted by the contractor showed adequate safety factors. The Ministry, however, refused to substitute wire mesh for the steel strips of the original RECO design. The contractor then proposed to use steel strips with wire ribs welded to the strip surface to resemble the conventional ribbed strips used in the reinforced earth system. Extensive pullout tests were performed at Utah State University on both conventional RECO ribbed strips and the new proposed strips. Both strips were embedded in a silty coarse sand backfill material and tested under overburden pressures corresponding to approximately 6 m (20 ft), 12 m (40 ft), and 18 m (60 ft) of fill. The soil was compacted to 95 percent of maximum density as per AASHTO T-99 Method C and allowed to reach equilibrium under the vertical load for at least 30 min before testing. The pullout resistance of the welded strips was approximately 15 percent higher than that of the conventional RECO strips; the welded ribs were not damaged or sheared off during testing.

Based on the testing results supplied, substitution was allowed, and the alternative scheme was constructed (Figure 4).

CONCLUSION AND RECOMMENDATIONS

Although the basic principle of the MSB walls is the same, there are key differences among the various systems that affect their stability and long-term performance. These differences are mainly in the soil-reinforcement interaction, the strength and stiffness of the reinforcement, the bond between soil and reinforcement, and the durability of the system. Other factors that make a difference include the strain compatibility between soil and reinforcement, the deformation characteristics of the backfill material, and the aesthetic and environmental impacts of the wall facing. These differences should be carefully evaluated when comparing two MSB systems.

Because of the influx of the MSB systems into the U.S. market and the serious differences among the many systems, the wall design should not be left freely in the hands of the contractor. Selecting the best system or proving an equal is, therefore, a difficult task facing the design engineer who must be familiar with the differences among the various systems and their impacts on the wall's performance. A preferred method that would ensure low cost, speedy process, and minimum confrontation would be to design a number of alternatives and include them in the bidding documents. The contractor is then asked to bid on one or more of the already-designed alternatives. In this way, the alternative designs will not be questionable and the procurement process will allow fair and equitable competition.

In all the case studies presented, regardless of the contracting procedures used, the MSB walls finally constructed were more economical than the other retaining walls included in the bidding documents.

ACKNOWLEDGMENTS

All four projects discussed were designed by Parsons Brinckerhoff under contracts with the Hawaii Department of Transportation, the Mass Transit Administration of the Maryland Department of Transportation, the New York School Construction Authority, and the Ministry of Public Works of the Hashemite Kingdom of Jordan. Internal stability analyses for alternative designs were performed by the Reinforced Earth Company, the Tensar Corporation, and the Hilfiker-Texas Corporation.

REFERENCES

1. Mitchell, J. K., and W. C. B. Villet. *NCHRP Report 290: Reinforcement of Earth Slopes and Embankments*. TRB, National Research Council, Washington, D.C., 1987.
2. *Code of Practice for Strengthened/Reinforced Soils and Other Fills*. BS 8006 Draft for Public Comments. British Standards Institution, 1991.
3. Christopher, B. R., S. A. Gills, J. R. Giround, I. Juran, J. K. Mitchell, F. Schosser, and J. Dunnicliff. *Design and Construction Guidelines for Reinforced Soil Structures—Volume 1*. Report FHWA-RD-89-043. FHWA, U.S. Department of Transportation, 1990.
4. Mitchell, J. K., and B. R. Christopher. North American Practice in Reinforced Soil Systems. In *Design and Performance of Earth Retaining Structures*. Geotechnical Special Technical Publication 25, American Society of Civil Engineers, 1990.
5. Schlosser, F., and V. Elias. Friction in Reinforced Earth. Presented at ASCE Symposium on Earth Reinforcement, Pittsburgh, Pa., 1979.
6. Lawson, C. R. Soil Reinforcement with Geosynthetics. *Proc., Workshop on Applied Ground Improvement Techniques*, Southeast Asian Geotechnical Society, Asian Institute of Technology, Bangkok, Thailand, 1992.
7. Munfakh, G. A. Innovative Earth Retaining Structures: Selection, Design and Performance. In *Design and Performance of Earth Retaining Structures*. Geotechnical Special Technical Publication 25, American Society of Civil Engineers, 1990.
8. Castelli, R. J., and G. A. Munfakh. Geotextile Walls in Mountain Terrain. Presented at 3rd International Conference on Geotextiles, Vienna, Austria, 1986.

Publication of this paper sponsored by Committee on Geosynthetics.

PART 2

Properties of Geosynthetics and Geocomposites

Survivability and Durability of Geotextiles Buried in Glenwood Canyon Wall

J. R. Bell and Robert K. Barrett

Geotextiles buried for up to 11 years in a geotextile-reinforced soil retaining wall constructed in 1982 in Glenwood Canyon, Colorado, by the Colorado Department of Transportation were exhumed from the wall in 1984 and again in 1993. Survivability and durability of the geotextiles were evaluated by comparing the wide-width tensile strengths of the excavated samples to the strengths measured before construction. The geotextile-reinforced wall was built by conventional methods with a very coarse, rounded, well-graded, pit-run gravel as the backfill soil. Four nonwoven geotextiles in two weights each were included in the wall. Wide-width tensile tests were performed on 31 exhumed samples of eight specimens each, resulting in 248 tests. Sample mean strengths were compared with preconstruction mean strengths. The results showed that exhumed sample strengths were lower by 4 percent to 51 percent. The average mean strength loss was 27 percent. For the conditions of this wall, construction was the dominate cause of damage. Little if any degradation occurred during the 9 years between the first and second sampling. The large sizes of the cuts and abraded areas in the exhumed geotextiles made small specimen tests, such as the burst or grab tensile tests, impractical. Some conclusions were limited by the large coefficients of variation for some damaged specimen populations, which required samples of more than eight specimens for reasonable precision.

Since their first use, there have been concerns about the survivability and durability of geotextiles. Are they damaged by construction? Are they degraded by long-term burial? This paper presents the results of a study by the Colorado Department of Transportation that provides some answers to these questions.

In spring 1982, an experimental geotextile-reinforced soil retaining wall was constructed in Glenwood Canyon, Colorado, as part of the Interstate 70 project. Four relatively low-strength nonwoven geotextiles in two weights each were included in the wall. The wall and its performance have been described elsewhere (1). The wall was to facilitate construction and was temporary. It was, therefore, decided to exhume geotextile samples from the wall after its design life and compare their strengths with the initial strengths of the geotextiles. The excavations were performed in two phases. The first was 2 years after construction in summer 1984 and the second was 11 years after construction in 1993.

The wall construction was the conventional U.S. Forest Service wrapped-face method. The backfill was end dumped on the geotextile, spread with a small bulldozer, and compacted by a vibratory smooth drum roller. The backfill was a free-draining, pit-run, rounded, well-graded, clean, very coarse sandy gravel. Nearly 100 percent was smaller than 150 mm (6 in.) with about 50 percent larger than 20 mm (0.75 in.) and 30 percent passing the No. 4 U.S.

standard sieve. Construction specifications required compaction to 95 percent of AASHTO T-180. The wall was 4.5 m (15 ft) high and 100 m (330 ft) long. A typical section is indicated in Figure 1.

The nonwoven geotextiles used are described in Table 1. The designations are appropriate for 1982 when the wall was built. The project was divided along its length into 10 segments, each 10 m (33 ft) long, and only one geotextile type. In some segments, the top nine layers contained the lighter weight fabric and the lower layers the heavier weight.

SAMPLING AND TESTING

The scheme was to investigate the effects of burial in the wall by comparing preconstruction wide-width tensile strengths to the strengths of exhumed samples. It was reasoned that the results could be influenced by

- Duration of burial;
- Geotextile type, polymer, and weight;
- Fabric variability;
- Construction stresses;
- Wall stresses due to gravity and loads;
- Damage during excavation and storage; and
- Test methods and procedure.

To address each of these factors, it was planned to

- Sample at least two times after construction;
- Sample each geotextile and weight each time;
- Sample at several locations within the wall section;
- Test eight randomly selected specimens per sample; and
- Always follow the same procedures for excavation, storage, and testing.

Sampling

Two years after construction (1984), samples were taken to investigate survivability. At that time, degradation due to aging was assumed small, and strength loss was attributed to construction, postconstruction traffic, and internal wall stresses from gravity. The 1993 samples were taken 9 years after the first samples so aging effects could have become apparent. The study was limited by time and other constraints that made it impossible to sample all fabrics both times.

Samples were planned from five locations in the wall section, as indicated in Figure 1. Different depths were chosen to show the effects of overburden pressure. Layer 3 was the highest layer that

J. R. Bell, Department of Civil Engineering, Oregon State University, Corvallis, Oreg. 97331. Robert K. Barrett, Colorado Department of Transportation, Grand Junction, Colo. 81502.

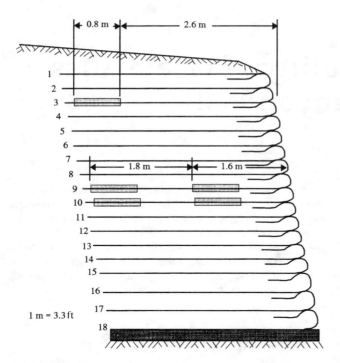

FIGURE 1   Glenwood Canyon wall section and sample locations.

contained all geotextile types and weights. In some wall segments, Layers 1 and 2 contained odd fabrics. Layer 9 was the lowest layer that contained all fabrics and weights. Layer 10 was the layer closest to Layer 9 that contained the heavier fabric in wall segments with two fabric weights.

In Layer 3, samples were planned from a zone well back from the wall face. At this location, outside of the theoretical Rankine active zone, shear stresses from the wall are nil. Also at this depth, overburden stresses are low, and the main effects should have been from construction stresses and postconstruction traffic. In Layers 9 and 10, samples were removed from zones both near the wall face and well back from the face. Back from the face, effects would be from overburden and construction; traffic and shear effects would be negligible. Near the face, shear stresses in the Rankine active wedge may also have affected the fabrics. Samples were identified by wall segment number, layer number, and whether from the front or back. For example, Sample 5-9F would be from the front of Layer 9 in wall Segment 5.

This sampling plan was strictly adhered to in 1984 except in Segment 6; Layer 11 instead of 10 was sampled as a result of misalignment of the geotextile seam in Layer 10 during construction. In 1993, because of time and other constraints, sheets were excavated from Layers 1 and 2 instead of Layer 3, and from Layer 8 instead of Layer 9. Samples from Layer 1 and some samples from Layer 2 were taken in the center of the sheet and are designated with a C.

The general excavation procedure was to dig a pit straddling the sewn seam between two desired wall segments. The pit was wide enough to give a fabric strip at least 750 mm (30 in.) wide on each side of the seam. Thus, two different fabric types were represented in each excavation pit. Power equipment was used to advance the pit, but hand methods and great care were used to remove the last soil layer above the geotextile sheet to be sampled. Any observed excavation damage was indicated on the geotextile with a marker pen. The sheets were labeled, the back and top marked, and placed in opaque plastic bags for storage and shipping.

Specimen Selection

Each test sample consisted of eight specimens 203 mm (8 in.) by 203 mm (8 in.). The scheme for selecting specimens in a sample is

TABLE 1   Test Geotextiles and Unaged (1982) Parameters

"Trade Name" (Manufacturer) Code No.	Nominal Mass g/m^2	Filament (Construction) Polymer	Unaged Strength kN/m	Failure Strain %
"Trevira" (Hoechst Fibers)		Continuous (Needled)		
H1115	170	Polyester	6.8	80
H1127	370		16.6	75
"Fibretex" (Crown Zellerbach)		Continuous (Needled)		
CZ200	200	Polypropylene	5.8	140
CZ400	400		10.1	145
"Supac" (Phillips Fibers)		Staple (Needled &		
P4oz	135	Heat Bonded*)	12.3	65
P6oz	200	Polypropylene	24.3	60
"Typar" (DuPont)		Continuous (Heat Bonded)		
D3401	135	Polypropylene	7.7	60
D3601	200		12.6	55

* One side only   1 g/m^2 = 0.03 oz./yd.^2   1 kN/m = 5.7 lb./in.

illustrated in Figure 2. Areas 610 mm (24 in.) wide by 813 mm (32 in.) along the seam were laid out on the geotextile sheets on each side of the seam as shown. For Layers 2 and 3, the two areas at the back were used. For the deeper layers, all four areas were used. Sheets from Layer 1 and some from Layer 2 were taken from the middle of the wall segments and, therefore, included only one geotextile type and did not include a seam. On these sheets a single area near the center of the exhumed geotextile sheet was used.

Eight specimens were selected from the 12 possible in each area by a blind draw. The selected specimens were cut from the sheet, labeled, and marked to show their orientation with respect to the wall. Specimens were tested regardless of fabric damage. Adjustments were made only if damage marked as due to excavation occurred in the specimen test area. In this event, a substitute specimen was cut from the nearest available location. A total of 248 wide-width tensile strength specimens from 31 samples cut from 21 exhumed geotextile sheets were tested.

Test Procedure

At the time of the preconstruction testing in 1982 there was no American standard wide-width tensile strength test method; however, except for the grips, the method used (2) was the same as ASTM D4595 approved in 1986. All tests in 1984 and 1993 were performed by the same procedure and with the same grips used in 1982.

The full width of the 203-mm (8-in.) by 203-mm (8-in.) specimen was held by the test grips, as illustrated in Figure 3. The specimens were orientated to measure the strength perpendicular to the wall face (cross-machine direction). The initial grip spacing was 102 mm (4 in.). The specimens were placed in the grips to test the middle 102 mm (4 in.) without regard to specimen damage. The geotextile specimens were conditioned by soaking in water for a minimum of 12 hr before testing. The tests were performed at a constant deformation rate of 10.2 mm/min (10 percent per min.) in a MTS Systems Corporation test machine. Load and elongation outputs were recorded by an *x-y* plotter.

The loads on the geotextiles in force per unit width (kPa/m) at various strains (percentage) were computed from the *x-y* plots and tabulated for each specimen. The maximum, minimum, and mean were determined at each of several strains for a sample. For

FIGURE 3 Geotextile specimen with major damage in test grips.

illustration, plots of these data for 1984 tests Typar 3601 Sample 1-3B are plotted in Figure 4 with the corresponding initial 1982 strength curves.

RESULTS

Plots such as Figure 4 are interesting and useful, but where a large number of samples are involved they are cumbersome. To simplify the results presentation, each load-strain curve is represented by the strength defined as the maximum load per unit width. The results in this form are summarized in Table 2. This table also presents failure strains and, for the mean strengths standard deviations, coefficient of variance and retained strength ratio. Coefficient of variance is defined as the ratio of the sample standard deviation to the mean and is expressed as a percentage.

The retained strength ratio is used as the measure of survivability and durability and is defined as the ratio of the mean sample strength from Table 2 to the initial (1982) mean strength from Table 1. As an example, the retained strength ratio value for Typar D3601 Sample 1-3B illustrated in Figure 4 is 65 percent. Retained strength ratio values for most samples are presented in Table 3.

Also listed in Table 2 is the number of specimens required in a sample to ensure that the mean of the sample tested represents the true mean of the geotextile sheet with an accuracy of plus or minus 10 percent with probabilities of 90 and 95 percent as calculated by the methods of ASTM D-2905. It is apparent from this table that less than half of the samples have a greater than 95 percent probability of 10 percent accuracy, and only about two-thirds have better than 90 percent probabilities of this accuracy. In the worst case, the probability of 10 percent accuracy is only about 70 percent.

DISCUSSION OF RESULTS

All the excavated samples have lower mean peak strengths than the original geotextile samples except the 1984 Fibretex 400 (CZ400) samples, which are higher. The Fibretex data are inconsistent and irrational. There is no known explanation for this inconsistency. This fabric was not sampled in 1993; therefore, although the results are included in Table 2 for completeness and to illustrate a sampling problem, further discussions ignore the CZ400 tests.

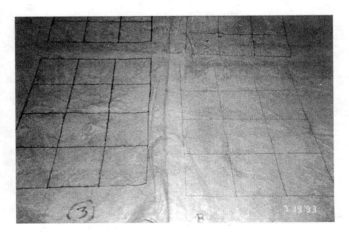

FIGURE 2 Exhumed geotextile sheet with specimen locations marked.

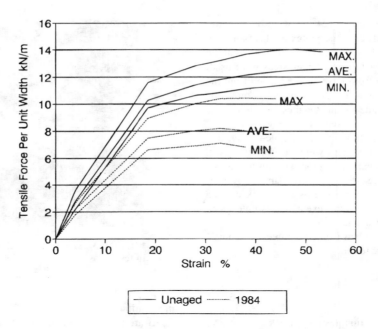

FIGURE 4   Load versus strain for D3601 1984 Sample 1-3B and unaged tests.

Layer 1 is a special case. This was a cover layer and not structurally part of the wall; therefore, the same care to protect the fabric during construction was not exercised for this layer as the others. Also, because it is the top layer, construction and postconstruction traffic may have caused greater damage to this layer than to Layers 2 and 3. Therefore, the low retained-strength ratios for Layer 1 are not considered comparable to the values from the other layers. They are included in Table 3 for D3401 and H1115 geotextiles to show the

relative magnitudes of damage that did occur in some instances. The Layer 1 values are shown in parentheses in Table 3 and are not included in the averages. Layer 1 values for CZ200 and P6oz fabrics are omitted from Table 3 for simplicity and are also not included in the averages. These considerations eliminate CZ200, CZ400, and P6oz from further discussions.

The average retained strength ratio as shown in Table 4 for the samples represented in Table 3 is 73 percent and the range is 49 per-

TABLE 2   Summary of 1984 Test Results

Fabric	Sample	Strength						Required* Specimens		Mean Strain
		Max. kN/m	Min. kN/m	Meam kN/m	Retained %	Std.Dev. kN/m	Coef.Var. %	95%	90%	%
Tervira H1115	6-3B	5.2	3.6	4.4	65	0.6	14	11	8	55
	6-9B	5.7	3.6	5.1	75	0.7	15	12	8	59
	6-9F	6.1	4.5	5.2	76	0.6	12	8	6	60
Tervira H1127	2-3B	10.5	7.8	9.5	57	0.9	10	6	4	57
	2-9B	9.7	8.5	8.9	54	0.4	5	3	1	61
	2-9F	9.6	6.8	8.1	49	0.8	9	6	4	59
	6-11B/F	12.6	6.9	9.7	59	1.8	19	19	14	50
Fibretex CZ400	4-3B	17.6	10.5	13.8	137	2.9	21	24	17	141
	4-9B	16.5	10.9	14.0	139	2.2	16	14	10	140
	4-9F	14.7	10.6	13.3	132	1.4	11	7	6	150
Supac P4oz.	7-3B	13.0	9.2	10.9	89	1.3	12	8	6	44
	7-9B	11.3	6.8	9.2	75	1.5	16	14	10	58
	7-9F	13.4	10.4	11.7	96	1.2	10	6	4	55
Typar D3401	5-3B	6.4	4.8	5.7	74	0.5	9	6	4	39
	5-9B	6.2	4.5	5.4	71	0.6	12	8	6	32
	5-9F	7.7	5.4	6.5	85	0.9	13	10	7	53
Typar D3601	1-3B	10.5	7.0	8.1	65	1.1	14	11	8	31
	1-9B	11.4	9.8	10.6	85	0.6	5	3	1	42
	1-9F	13.0	8.7	10.7	85	1.3	12	8	6	36
	5-10B	11.9	9.8	11.1	89	0.7	6	3	1	38
	5-10F	12.4	9.8	10.8	86	0.8	8	4	3	42

* Number to give indicated probability of 10% accuracy.   1 kN/m = 5.7 lb./in.

TABLE 3 Summary of 1993 Test Results

Fabric	Sample	Strength						Required* Specimens		Mean Strain
		Max. kN/m	Min. kN/m	Mean kN/m	Retained %	Std.Dev. kN/m	Coef.Var. %	95%	90%	%
Tervira H1115	6-1C	4.1	2.0	2.8	41	0.7	25	35	25	56
Tervira H1127	10-2B	11.5	7.6	10.0	60	1.6	16	14	10	61
	10-8B	11.9	9.7	11.3	68	0.7	6	3	1	59
	10-8F	11.1	7.6	9.5	58	1.1	12	8	6	59
Fibretex CZ200	9-1C	5.5	3.9	4.7	80	0.6	13	10	7	122
Supac P6oz.	10-1C	23.4	19.8	21.1	87	1.3	6	3	1	54
Typar D3401	8-1C	4.8	3.5	4.0	53	0.4	11	7	6	23
Typar D3601	9-2B	10.4	7.3	9.0	72	1.1	12	8	6	31
	9-8B	12.1	9.4	10.5	83	1.0	9	6	4	38
	9-8F	11.2	4.0	8.7	69	2.4	28	42	29	31

* Number to give 90% or 95% probability of 10% accuracy.   1 kN/m = 5.7 lb./in.

cent to 96 percent. Also, failure strains are generally lower for the exhumed samples. It is important to note that the percentage reduction in peak strength may be either more or less than the percentage reduction in stress at low strains. The general trend is for the stress reduction to be less at low strains; therefore, the interpretations made in this paper may be conservative when considered relative to working stresses.

Only one backfill soil was used in the wall and, although this material was not the worst that could have been selected, it probably was more damaging to the geotextiles than most backfills would have been. The large particle sizes concentrated stresses and a geotextile directly between two large particles might have suffered greater damage than if the backfill had been finer. The material was, however, well graded and the particles were rounded. Compaction was greater than usually specified, which could have contributed to geotextile damage.

Only one construction procedure and one set of equipment were used. The construction methods were conventional. Greater care

may have reduced geotextile damage but would probably not have been cost effective.

Figure 3 shows a geotextile test specimen with major visible damage. Nearly all specimens had some visible damage but there were great variations. Sometimes there was only slight abrasion. Sometimes there were cuts and tears more than 20 mm (0.75 in.) long, as in Figure 3. Some specimens had several visible cuts and some had none. This resulted in high sample standard deviations and reduced accuracy, making it impossible, with the number of specimens tested, to identify minor effects or make fine distinctions between factors. Only relatively large differences are statistically significant.

Early in the planning of the study, it had been anticipated that relatively few wide-width tensile strength tests could be used to measure the retained strength and a large number of burst tests could be used to evaluate variability. The large sizes of many of the damaged areas made the burst test impractical. This required many more large specimen wide-width tensile tests, increased the cost, and reduced the total number of tests possible.

TABLE 4 Retained Strength Ratio Values

	Layer and Location	Year Sampled	Retained Strength Ratio (%)				
			Geotextile				
			D3601	D3401	H1127	H1115	P4oz
Upper Layer Values	1C	1993		(53)		(41)	
	2B	1993	72		60		
	3B	1984	65	74	57	65	89
Lower Layer Values	8B	1993	83		68		
	8F	1993	69		58		
	9B	1984	85	71	54	75	75
	9F	1984	85	85	49	76	96
	10B	1984	89				
	10F	1984	86				
	11B/F	1984			59		
Average* Values	Upper		69	74	59	65	89
	Lower		83	78	58	76	86
	1984		82	77	55	72	87
	1993		75		62		
	All		79	77	58	72	87
	Heat Bonded Polypropylene Samples		79				
	Needle-punched Polypropylene Samples		87				
	Needle-punched Polyester Samples		62				
	All Samples		73				

Values in ( ) not included in averages.

With the data available, comparisons may be attempted for the following factors:

- Duration of burial,
- Location in a layer (front or back),
- Depth in the wall,
- Geotextile mass,
- Geotextile construction, and
- Geotextile polymer.

Duration of Burial

Figure 5 presents the average retained strength ratios for the exhumed geotextiles by year sampled. There is no trend of increased damage with time of burial for the two fabrics sampled both years. Comparing the data in Table 3 for front versus back and upper versus lower leads to the same conclusion. Chemical tests are in progress and may show some time effects, but long-term durability as indicated by wide-width tensile strength is not a problem for the 9-year period between tests and for the conditions of this wall would probably not be a significant factor for any reasonable design life.

It is concluded that durability is not a factor in this wall. All loss of strength is due to construction and postconstruction traffic, which for this wall had ceased by 1984. Survivability, however, with sample strength reductions of up to 50 percent, is important and must be considered in design with appropriate partial factors of safety.

Since durability is not a factor, 1984 and 1993 test results are combined to increase the data base for all further comparisons.

Location in Layer

Excavated geotextile sheets from the lower layers were sampled front and back. Table 3 shows eight pairs of samples. These are presented graphically in Figure 6. Of the eight, three have nearly the same retained strength ratios, three have greater, and two have lower retained strength ratios for the back samples. When all pairs are averaged, the ratio of front-to-back strength ratios is nearly one. There is indication that there may be somewhat greater damage to the geotextiles near the face of the wall relative to the fabrics well back from the face and that the damage is progressive. This would suggest the shear stresses in the Rankine active wedge contribute to the fabric damage, but considering the variations of the samples the evidence is not persuasive.

Depth in Wall

Table 4 separates the geotextiles in the upper part of the wall (Layers 1, 2, and 3) from those in the lower part (Layers 8, 9, 10, and 11). The averages of the retained strength ratios for upper, lower, and all samples for each geotextile are plotted on Figure 7. Of the five geotextiles for which there are data for both zones, two indicate what may be significantly greater damage in the upper layers and three show no significant difference. This supports the conclusion that damage is due to construction with some additional damage by postconstruction traffic in the upper layers. There is no indication, for the depth investigated, that the weight of the overlying material contributes to the geosynthetic damage.

Geotextile Mass

Two heavier geotextiles (D3601 and H1127) are compared with the lighter weight fabrics of the same type (D3401 and H1115) in Figure 7. H1127 and H1115 show considerable difference, but comparison indicates the heavier fabric suffers greater relative damage. This is counterintuitive. Table 3 shows that there are only three samples for H1115, and the overall average for this fabric is strongly influenced by the two lower samples, which have the highest retained strength ratios of all the 10 Trevira samples. It appears

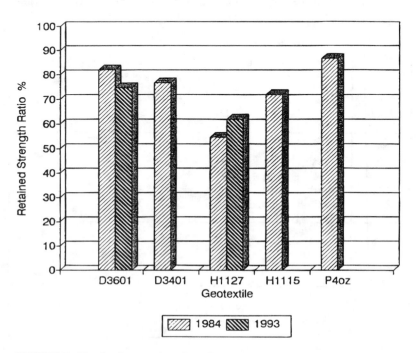

FIGURE 5 Retained strength ratio and year sampled.

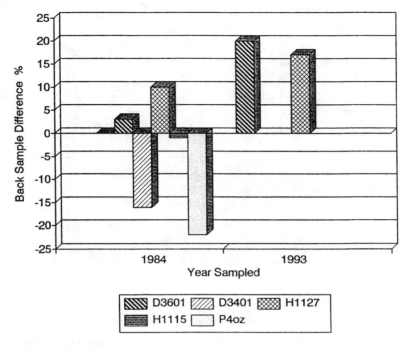

FIGURE 6 Retained strength of back samples relative to front samples.

that for the test conditions and the geotextiles used, little if any variation exists in the relative damage of different weights of the same geotextile.

Geotextile Construction and Polymer

The averages of all retained strength ratios for the three types of geotextiles represented by the data are presented in Figure 8. These are a heat-bonded polypropylene, a needle-punched polyester, and a needle-punched polypropylene. The first two are each represented in two weights. Only one weight of needle-punched polypropylene was tested. This needled polypropylene (Sumac P4oz) is constructed of staple filaments; the others have continuous filaments. This geotextile is also lightly heat bonded on one side. This needled polypropylene is represented by only four samples (32 specimens). There are at least 10 samples (80 specimens) each for the other two.

The polypropylene geotextiles samples retained an average of 81 percent of their initial strengths, and the polyester fabrics retained an average of 62 percent. The needled polypropylene may suffer slightly less relative damage than the heat-bonded polypropylene, but there are too few samples to consider this difference significant, so all polypropylene samples are considered in the above average. These data suggest partial factors of safety of 1.25 and 1.6 for nonwoven polypropylene and polyester geotextiles, respectively. It appears reasonable that partial factors of safety for survivability should be different for different polymers because they have different mechanical characteristics.

Summary

The greatest damage is mechanical abrasion and cutting due to construction operations. At least for the duration of this study, there is no significant decrease in strength with time after the first excavations to indicate chemical aging or continued degradation from in situ stresses. There is some indication of reduced strength in the upper layers that may be from postconstruction traffic.

CONCLUSIONS AND RECOMMENDATIONS

The study included a coincident series of 10 geotextile reinforced earth-retaining walls. Each was 10 m (33 ft) wide and 4.5 m high, and all were constructed with coarse, rounded, well-graded pit-run gravel backfill and with a variety of relatively low-strength, nonwoven geotextile reinforcements. The wall was constructed using the traditional U.S. Forest Service wrapped-face methodology. The backfill was compacted to at least 95 percent of AASHTO T-180 with a large vibratory smooth drum roller. The test walls were faced with shotcrete 3 months following construction.

Viable survivability and durability data were obtained from three nonwoven geotextiles. Survivability and durability were evaluated by comparing wide-width tensile strength of samples exhumed in 1984 and 1993 with initial strengths measured before construction in 1982. Conclusions are limited to these specific conditions and geotextiles.

- There was no loss of strength in samples obtained and tested in 1993 compared with 1984. Durability was not a problem in this wall.
- There was an average loss of strength for all samples of 27 percent, principally as a result of construction damage. Survivability was a significant factor for this wall.
- Not all geotextile were equal in construction survivability. The polypropylene geotextile samples lost an average of 19 percent of their strength to construction damage, and the polyester geotextile samples lost an average of 38 percent.

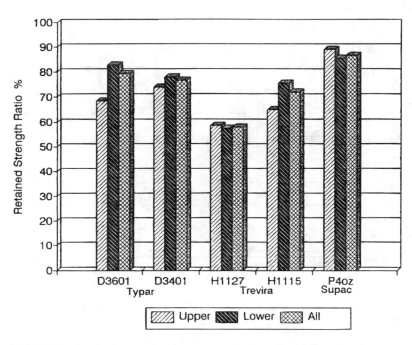

FIGURE 7   Retained strength ratio versus vertical location in the wall and mass.

• Choice of backfill and construction methods and equipment contributed to the initial loss of strength. These could be considered moderately severe conditions.

• There was little difference in relative strength loss between lighter weight and heavier weight fabrics of the same type and polymer.

• The large coefficients of variance for the damaged specimen populations required relatively large samples to yield reasonable accuracy. Further, the large sizes of the cut and abraded areas within the specimens eliminated the use of index tests, such as the burst or grab tensile tests.

This study provides preliminary design parameters for the use of nonwoven geotextiles in moderately severe construction conditions. The study shows that in situ stresses and aging did not contribute significantly to the degradation of the geotextiles. It is concluded

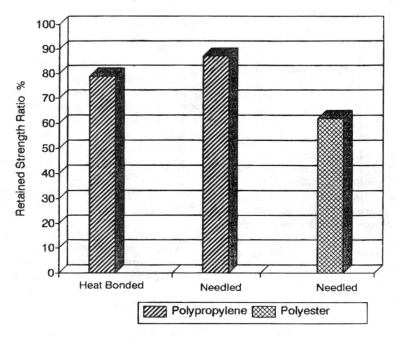

FIGURE 8   Retained strength ratio versus geotextile construction and polymer.

that the partial factors of safety for survivability of 1.25 for polypropylene and 1.6 for polyester may be used as conservative values for most reinforced walls constructed with 140 to 400 g/m² (4 to 8 oz) nonwoven geotextiles. These values are probably over-conservative for sand backfills, but for large angular crushed rock backfills damage could easily exceed these recommendations.

Because of the many possible combinations of backfills and reinforcements and because of diverse construction specifications and equipment, extensive field testing will be required before confidence is gained in interpolated and extrapolated survivability values.

As a closing recommendation, each department of transportation and other agencies using geotextile reinforcement applications is urged to start developing suites of data for the typical backfills, construction methods, and choices of reinforcements. These data could be obtained most cost effectively through exhumations during actual construction projects, but preconstruction evaluations with test backfills approximating actual construction conditions and methods may be justified on large or critical projects, or both. Testing should be directed by the designer of record. Particular care must be exercised in exhuming the samples. To be most valuable, the results of these studies must be published.

ACKNOWLEDGMENTS

The authors gratefully acknowledge the Colorado Department of Transportation for financial support of all phases of the study. The Exxon Corporation and Hoechst Celanese Corporation provided financial support for the 1993 phase of the study, and Joel Sprague of Sprague and Sprague Consulting Engineers and Richard Goodrum of the Hoechst Celanese Corporation supervised the 1993 excavations.

REFERENCES

1. Bell, J. R., R. K. Barrett, and A. C. Ruckman. Geotextile Earth-Reinforced Retaining Wall Tests: Glenwood Canyon, Colorado. In *Transportation Research Record 916*, TRB, National Research Council, Washington, D.C., 1983, pp. 59–69.
2. Shrestha, S. C., and J. R. Bell. A Wide Strip Tensile Test of Geotextiles. *Proc., 2nd International Conference on Geotextiles*, Vol. 3, Las Vegas, Nev., 1982, pp. 739–745.

Publication of this paper sponsored by Committee on Geosynthetics.

Pullout Mechanism of Geogrids Under Confinement by Sandy and Clayey Soils

DAVE TA-TEH CHANG, TSUNG-SHENG SUN, AND FAN-YI HUNG

The frictional mechanism of geogrid-soil interaction is considered in two parts: one is passive resistance from the soil mass ahead of the transverse ribs and the other is produced by the frictional resistance. The factors influencing the mechanical performance of geogrid-soil interaction are studied. With three types of geogrids, three types of soils (two sandy soils and one clayey soil), and various testing conditions, a series of pullout tests for geogrid was conducted. Through the testing program, the significance of influencing factors and the strain contribution measured by strain gauges are studied. According to the results of the experiments, the pullout resistance of geogrid tends to increase as the confining pressure is increased. For sandy soils, the passive earth pressure offers the most pullout resistance; when using fine grained soil, it is replaced by friction resistance.

The concept of earth reinforcement involves placing certain materials in the soil to increase the bearing capacity of the soil mass and to stabilize slopes. From the civil engineering standpoint, excessive deformation of the reinforced soil mass is prevented by frictional resistance that occurs where the soil grains are in contact with the reinforcement element.

Geogrid is effective as a reinforcement element because it offers the following two forms of resistance to the pullout failure mechanism when used as a reinforcement element for soil structures: (a) friction between soil and the surface of the geogrid and (b) the passive earth resistance of the soil against the transverse ribs. The researchers' investigations have been focused on ascertaining which of these two offers the greater resistance in the geogrid-soil interaction.

To obtain rational parameters for design, it was taken into consideration that soil available on the work site is generally the backfill material of choice, owing to the difficulty of obtaining sand for use in public construction projects. For this reason a reinforced earth wall demonstration site was established for this study in Tianliao, a mudstone district in Kaohsiung County on the route of a new freeway system. Two types of geogrid were used to examine and compare pullout interaction behavior with sandy and clayey soils.

MECHANISMS OF INTERACTION BETWEEN REINFORCEMENT AND SOIL

Stress transfer between the geogrid and the soil is primarily a function of frictional resistance and passive earth pressure. The former is generated by friction between the soil and the surface of the geogrid; the latter is a function of the grid-shaped construction, which causes the transverse ribs to interlock tightly with the inter-

vening soil. Because pullout force causes the geogrid to move relative to the surrounding soil, passive earth pressure develops against the transverse ribs. Thus a reinforcing effect is attained by bringing into play the latent interaction potential inherent in friction resistance and soil passive resistance.

For pullout-resistance behavior, many properties of soil and geogrid are known as the influence factors (*1*). The pullout resistance to geogrid is thought to be developed by the following two stress-transfer mechanisms: (a) frictional resistance between soil grains and contact grid surface and (b) the resistance from the soil passive mass against the transverse ribs (*2*). To evaluate the frictional resistance (P_f), and ideal expression has been derived and suggested (*3*):

$$P_f = 2A_r\alpha_s \cdot \sigma_v \cdot \tan \delta \qquad (1)$$

where

   A_r = gross area of geogrid,
   α_s = fraction of solid surface area in grid,
   σ_v = vertical effective stress, and
   δ = friction angle between soil and geogrid surface.

Equation 1, the formula for estimating P_f, is widely accepted and frequently used. The passive resistance of soil bearing on the transverse ribs is a problem similar in kind to the base pressure on deep foundations in soil. Passive resistance is a function of the grid-shaped structure, which causes the transverse ribs to bind tightly with the intervening soil. Because pullout forces cause the geogrid to move relative to the surrounding soil, passive earth pressure develops against the transverse ribs. It has been suggested that this passive earth pressure be expressed in terms of the bearing capacity from the punching failure mode as given in Figure 1 (*2*).

To determine the potential interaction of the geogrid with not only granular soil but also fine-grained soils, a testing device was designed and built for this study. The device was used to ascertain the basic mechanical characteristics of the geogrid during pullout testing under confined conditions. Strain distribution along the geogrid was measured only to supply supplementary data; it is not a focus of this study. The confining box suggested by the Geosynthetic Research Institute (*4*) was not used in this study.

LABORATORY TEST PROGRAM

Materials

Two types of sandy soil were used in the test program: one was collected from the backfill sand used for the test wall in Tianliao,

Civil Engineering Department, Chung Yuan University, Chung Li, Taiwan 32023, R.O.C.

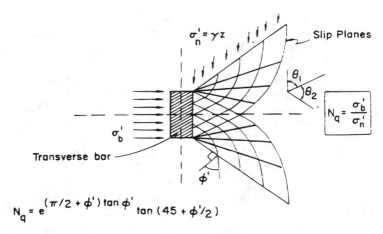

$$N_q = e^{(\pi/2 + \phi') \tan \phi'} \tan (45 + \phi'/2)$$

FIGURE 1 Passive bearing punching shear failure mechanism (2).

Kaohsiung County, and the other was Ottawa sand (C-190). Relative density for both was controlled at 80 percent. The clayey soil was weathered mudstone, which was also obtained from the same test site in Kaohsiung County. Water content was maintained at OMC + 2 percent, and the degree of compaction was 95 percent. See Table 1 and Figure 2 for the basic properties and grain-size distribution curves of the aforementioned soils. The two types of geogrids used were manufactured from HDPE, labeled "A" and "B" (Table 2). To reduce the boundary effect, samples of the A and B geogrids measuring eight squares in width and three squares in length were used. The portion of each geogrid buried in the soil was fixed at 39 cm. Another part of the study focused on exploring the frictional resistance of the geogrid-soil interface for specimens of the same size. Pullout tests were conducted on the A and B grids from which the transverse ribs had been cut. It is believed that pullout resistance from the trimmed specimen is a factor of frictional resistance only. In addition, the differences in contact area between trimmed specimens and intact specimens must be allowed for so that the frictional resistance for intact grid specimens can be calculated. The passive resistance from the transverse ribs, therefore, is determined for comparison purposes.

Pullout Box

The top and bottom pullout boxes are 40 cm long, 50 cm wide, and 15 cm deep internally, with a 1-cm opening between the two adjacent boxes. See the structural sketch in Figure 3. To prevent boundary effects from occurring, the upper and lower boxes were fitted with sleeves. Below the boxes is an adjustable bearing plate that allows the pulling forces to be aligned into the same plane as the geogrid, pulling it out via the opening between the two boxes. Normal stress is applied using air bags into which compressed air can

be directly pumped. Polystyrene packers are placed inside the air bag in the lower box to ensure both that the air compartment in the lower box is completely sealed during compaction of the sample, and that the normal stress is evenly distributed, reducing the effect of the laboratory boundary effect. Normal stress of 0.5, 1.0, and 1.5 kg/cm² is applied during the test. The pulling system consists of a constant rate motor assembly. The pulling force can be adjusted through a set of gears.

Measuring System

Two linear variable differential transformers (LVDTs) with a maximum stroke of ± 10 cm are attached to the puller. An amplifier accurate to 10^{-3} mm measures the pullout displacement of the geogrid, and the two LVDTs provide verification. The strain gauges are smeared with paraffin to prevent moisture-induced short circuits, and are cemented to the surface of the rib. In this way, one can ascertain the way strain is distributed when the geogrid is subject to pulling forces. The Kyowa KLM-6-A9 strain gauge with Kyowa EC-30 cement was used; with this combination, strains of up to 20 percent can be measured. The strain gauge is attached to three transverse ribs 2 cm from their junctions, ensuring that the transverse ribs are all the same width at the point of attachment. They are located 5, 21, and 37 cm from the front wall of the box. A Tokyo Sokki Kenkyujo TDS-301 data logger with an amplifier simultaneously records the values registered by load cells, LVDTs, and strain gauges.

Testing Procedures

During testing, strain rate was controlled at 1 mm/min (4); the portion of the grid buried in the soil was fixed at 39 cm. The soil in the

TABLE 1 Properties of Tested Soils

Property	Backfill sand	C-190 sand	Weathered mudstone
Dry unit weight, γ_d (g/cm³)	1.791	1.715	1.865
Angle of internal friction , ϕ	45 °	37 °	29 °
Cohesion, (kg/cm²)	------	------	0.364

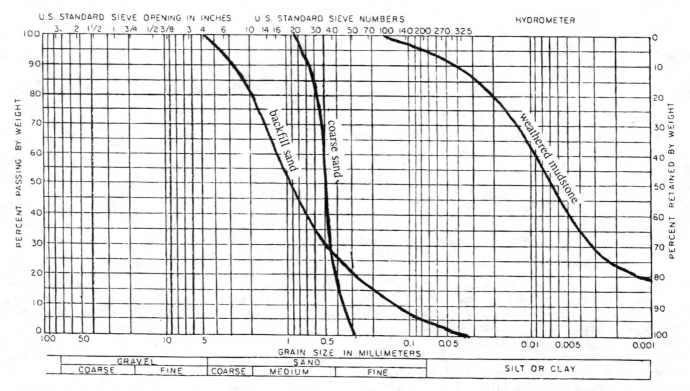

FIGURE 2   Grain size distribution of tested soils.

top and bottom pullout boxes was compacted into the boxes in five even layers, and the geogrid specimen was buried in the middle, aligned with the opening between the boxes and with the direction of pulling-force application. Compressed air was then pumped into the air bags in accordance with the required normal stresses and left for 24 hr so the pressure could equilibrate. The leads from the strain gauges were connected to the data logger, and when the reading from the strain gauge stabilized after 24 hr of pressurization—indicating that settlement of the sample had ceased—the LVDT was placed in position, the motor and data logger were turned on, and the data logger was set to take readings every 10 sec until the pullout forces decreased.

TABLE 2   Properties of Geogrids Used

Geogrid		A	B
Polymer type		HDPE	HDPE
Shape of apertures		oblong	oblong
Thickness of longitudinal ribs,	mm	1.4	0.95
Length of longitudinal ribs,	mm	144	144
Spacing of longitudinal ribs,	mm	16	16
Width of transverse ribs,	mm	16	16
Thickness of transverse ribs,	mm	3.9	2.7
Thickness of junction,	mm	3.9	2.7
Tensile strength,	kN/m	87	60
Elongation,	%	10.8	8.8

ANALYSIS AND DISCUSSION

As previously discussed, the two main mechanisms of pullout resistance for geogrids are the frictional resistance and the resistance due to passive earth pressure. However the pullout resistance exhibited by the geogrid varies with different soil media.

Effect of Soil Properties on Pullout Behavior of Geogrid

As Figures 4, 5, and 6 show, the geogrid's pullout resistance during initial pullout displacement in backfill sand is actually lower than in coarse sand or weathered mudstone. After completion of initial pullout displacement, pullout resistance rises steadily. This discrepancy arises because backfill sand is well graded and has a high proportion of large grains, which may slide easily when pressed by the transverse ribs. These large grains move until they are packed tightly against the smaller grains, giving rise to greater soil passive resistance and a concurrent steady increase in pullout resistance. The grains in C-190 sand and weathered mudstone have relatively lower φ values, so pullout resistance is more likely to stabilize at a constant value when the transverse ribs have caused the soil grains to slide.

To discover the relationship during the experiment between the friction resistance at the surface of the grid and the passive resistance against its transverse ribs, the pullout tests were conducted on A and B grids from which the transverse ribs had been removed. By adjusting the effective contact area of the geogrids in this way and ascertaining their frictional resistance with the soil, a comparison could be made between the frictional resistance figures of geogrids with and without transverse ribs. As shown in Figures 7 and 8, friction resistance makes up 30 percent of the total pullout resistance

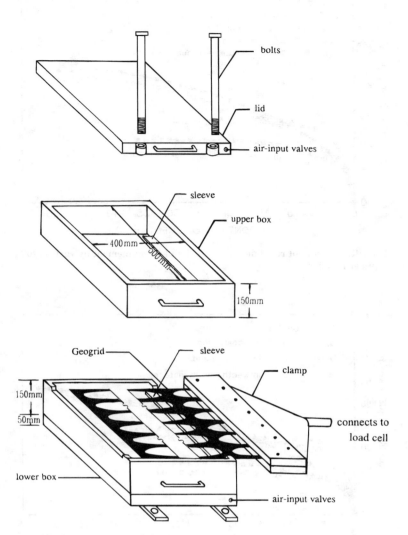

FIGURE 3   Confining pullout box.

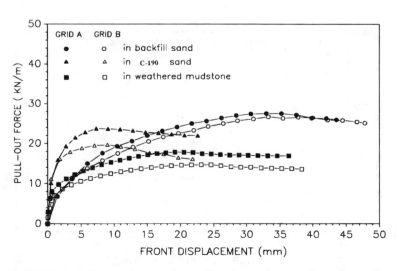

FIGURE 4   Pullout resistance of geogrid under confinement by various soil types at 0.5 kg/cm².

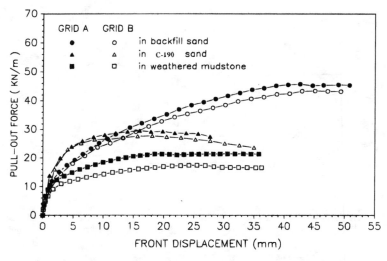

FIGURE 5 Pullout resistance of geogrid under confinement by various soil types at 1.0 kg/cm².

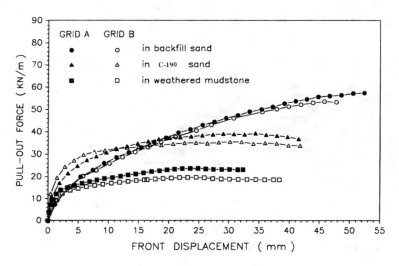

FIGURE 6 Pullout resistance of geogrid under various soil types at 1.5 kg/cm².

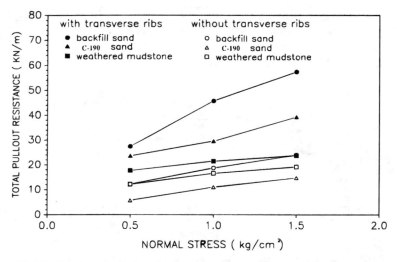

FIGURE 7 Comparison of pullout resistance in A geogrids with and without transverse ribs.

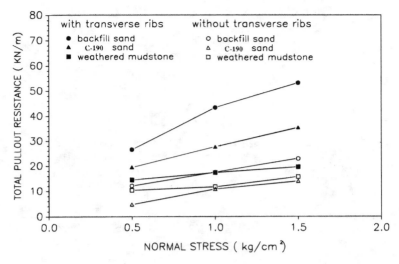

FIGURE 8   Comparison of pullout resistance in B geogrids with and without transverse ribs.

exhibited by the geogrid in sandy soil. The remaining 70 percent is contributed by soil passive resistance; thus, the effect of soil passive resistance is greater than that of frictional resistance. Numerical methods (5) were used to predict the components of pullout in dense sand: it was found that passive resistance made up most of the resistance. It can also be seen from Figure 9 that Grid A offers greater pullout resistance than Grid B. Although Grids A and B are almost identical in shape, the transverse ribs of Grid A are 1.2 mm thicker than those of Grid B, so the bearing surface for the soil passive resistance of Grid A is greater than that of Grid B; thus, the bearing surface of the transverse ribs directly influences the amount of soil passive resistance that is developed.

Some interesting findings were discovered in the weathered mudstone test results. A representative confinement finding of 1.5 kg/cm² is included in Figure 10. From Figure 10, it is observed that most of the pullout resistance exhibited by the geogrid in clayey

soils is contributed by frictional resistance; thus, the frictional resistance effect contributes more to the pullout resistance of the geogrid in weathered mudstone than does soil passive resistance. The findings also indicate that, during the initial stage when pullout displacement had not yet exceeded 2 mm, the total pullout resistance was equivalent to its friction resistance. It was only after this initial stage that total pullout resistance values gradually exceeded the frictional resistance figures. This phenomenon arises in the initial stages, where the displacement that occurs is due to elongation in the front portion of the grid specimen itself, not to relative displacement between the grid and the soil mass. The main resistance to pulling forces is contributed by static friction between the grid and the soil. The soil passive resistance of the transverse ribs develops only when there is relative displacement between the soil and the grid, so it is only when this displacement occurs that total pullout resistance gradually becomes higher than friction resistance.

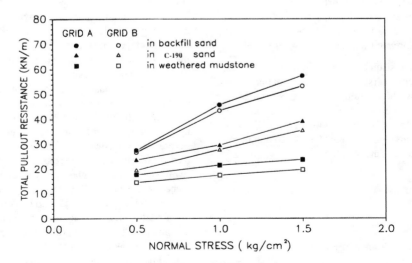

FIGURE 9   Total pullout resistance between geogrid and soil under various confinements.

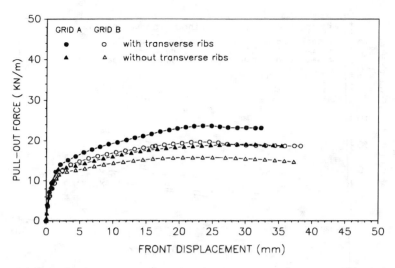

FIGURE 10   Pullout resistance relationships for geogrid under confining pressure of 1.5 kg/cm² in weathered mudstone.

From the evaluation described, it can be summed up that in sandy soils most of the pullout resistance is contributed by soil passive resistance. In clayey soils, frictional resistance provides the largest part of pullout resistance.

Bearing Resistance on Transverse Ribs

As has been pointed out, the pullout resistance is composed of frictional and passive resistance. This can be expressed with the following formula:

$$P_T = P_F + P_B \tag{2}$$

$$P_F = P_{f(LR)} + P_{f(TR)} \tag{3}$$

where

P_T = total pullout force,
P_F = total friction resistance at grid surface during pullout,
P_B = bearing force at transverse ribs,
$P_{f(LR)}$ = surface friction force at longitudinal ribs, and
$P_{f(TR)}$ = surface friction force at transverse ribs.

To calculate the bearing resistance generated at the transverse ribs in this experiment, the transverse ribs were cut from one of two identical geogrid specimens; then pullout tests were conducted for both. Because the resistance figures obtained were a function of the frictional resistance generated at the longitudinal ribs and the cut surface, the following modified formula must be used to obtain true total frictional resistance values during pullout:

$$P_F = \frac{A_{(TR + LR)}}{A_{(LR + \text{cut surface})}} \, P_{f(LR + \text{cut surface})} \tag{4}$$

where

$P_{f(LR + \text{cut surface})}$ = frictional resistance of longitudinal ribs and the cut surface,

$A_{(TR + LR)}$ = surface area of longitudinal and transverse ribs, and

$A_{(LR + \text{cut surface})}$ = surface area of longitudinal ribs and cut surface.

During pullout in sandy soils, the soil grains at the bearing surface of the transverse ribs are packed into a denser state, thus maximizing the interlocking effect between the soil and the transverse ribs and increasing the passive resistance at the bearing surface. On comparing the bearing forces at the transverse ribs under the effects of different confinements as depicted in Figure 11, it is evident that in backfill sand or coarse sand the bearing force rises as confining pressure is increased. This may be attributed to the densely packed structure of the grains in this well-graded backfill sand, which causes the rate of increase in bearing force to rise as more confining pressure is applied. The bearing force shows no marked increase in uniformly graded coarse sand however. The foregoing phenomenon explains why the angle of internal friction, the bearing area of the transverse ribs, and the vertical effective overburden stress all influence the bearing resistance at the transverse ribs. The passive bearing failure model (Figure 1) for geogrids in sandy soils is thus confirmed.

As described, the pullout behavior of the grid differs for sandy and clayey soils. It was discovered in this study that, under pullout action in clayey soils, a pullout failure plane was observed against the upper and lower surfaces of the longitudinal and transverse ribs of the grid. This is because the clay grains are very small and cohesive and also because the low angle of internal friction lessens shear resistance. This caused a "breakthrough" phenomenon that resulted from the knifelike cutting surface of the transverse ribs pressing against the soil in the grid's apertures during the pullout process. Figure 11 demonstrates that in clayey soils the bearing forces at the transverse ribs are not affected by confinement, and remain relatively stable. This explains why in clayey soils the passive resistance aspect of the pullout resistance is related to the degree of soil cohesiveness, and is not affected by confinement. Hence in the failure model depicted in Figure 1, there are no passive bearing zones; the only possibility is that failure is limited to elastic bearing zones. Thus, it can be seen that the formula of Jewell et al. (2) requires amendment, because it is not suitable for evaluating passive resis-

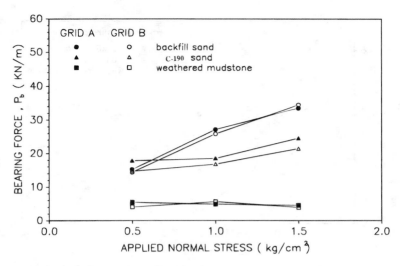

FIGURE 11   Comparison of bearing forces at the transverse ribs.

tance during pullout in cohesive soils. More data and further research are needed to establish the correct method.

Effect of Soil Confinement

Figure 9 shows the total pullout resistance of grids under normal stress in different soils. The pullout resistance of the grid increases with increasing confinement pressure, but the rate of increase differs depending on the soil type. Backfill sand gives higher rates of increase, and weathered mudstone gives the slowest rate of increase. Friction, the bearing capacity factors of soil passive resistance, and the angle of internal friction of the soil are all closely interrelated in terms of pullout resistance. Where the angle of internal friction of the soil is high, the bearing capacity factors will also be high, hence the relatively high rate of increase in pullout

resistance as confinement pressure is increased for grids in sand backfill. By contrast, the rate of increase in pullout resistance as confinement pressure is increased is relatively low for grids in weathered mudstone. Furthermore, because the thickness of the grid is significant, when pullout forces are applied, the displacement results in dilation of soil particles, which leads to an increase in confined pressure. Hence pullout resistance tends to increase.

Strain Distribution Along Geogrid

Figures 12 and 13 are strain-distribution diagrams for all measurements of monitoring points at which the pullout displacement is 25 mm. The diagram clearly shows that when pulling forces are applied to the grid, the greatest strains are found at the measurement points nearest the portion where the pulling force is being applied,

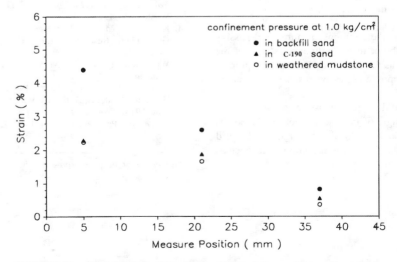

FIGURE 12   Strain distribution among all measuring points on Geogrid A when pullout displacement had reached 25 mm.

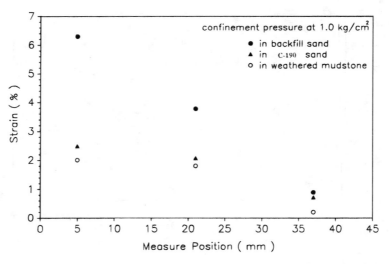

FIGURE 13   Strain distribution among all measuring points on Geogrid B when pullout displacement had reached 25 mm.

and most of the strain occurs at the two front measurement points. Thus most of the pullout resistance effect is provided by the grid's two front apertures. This unequal strain distribution under the pullout effect proves that strain transfer is uneven along the grid during pullout. This strain distribution pattern suggests that if the buried portion of the grid is too long strain will gradually be transferred from the front of the grid to the back when pulling forces are applied. This leads to the overdesign of the reinforcement material because when the front portion of the grid has achieved its maximum anchoring effect, the rear portion may not have undergone any deformation at all. This unequal strain-distribution phenomenon confirms the results obtained elsewhere (6).

CONCLUSION

Based on the foregoing, the following conclusions can be drawn from the analysis of the grid's mechanical characteristics.

• With the granular soil as the confining medium, soil passive resistance is the main contributor to the pullout resistance of the grid. With fine-grained soils as the confining medium, the proportion of pullout resistance composed of passive resistance decreased significantly.
• The pullout resistance of the grid increases as confining pressure increases. Where the angle of internal friction of the soil is high, the bearing capacity factors will also be high, so the grid exhibits a relatively faster rate of increase in pullout resistance as confining pressure is increased.

• For the grids used in the study, the pullout resistance in sandy soil was higher than in fine-grained soil.
• The strain distribution of the geogrid during testing was triangular; the strain gradually reduced from a maximum at the pullout end to zero at the other end.

REFERENCES

1. Bauer, G. E., and Y. M. Mowafy. The Effect of Grid Geometry and Aggregate Size on the Stress Transfer Mechanism. *Proc., 4th International Conference on Geotextiles, Geomembranes and Related Products*, Vol. 2, 1990, p. 801.
2. Jewell, R. A., G. W. E. Milligan, R. W. Sarsby, and D. Dubois. Interaction Between Soil and Geogrids. *Proc., Symposium on Polymer Grid Reinforcement in Civil Engineering*, Thomas Telford Limited, London, England, 1984, pp. 19–29.
3. Mitchell, J. K., and C. V. Villet. *NCHRP Report 290: Reinforcement of Earth Slopes and Embankments*. TRB, National Research Council, Washington, D.C., 1987.
4. Koerner, R. M. *GRI Test Methods and Standards: Geogrid Pullout.* Geosynthetic Research Institute, Drexel University, Philadelphia, Pa., 1992.
5. Wilson-Fahmy, R. F., R. M. Koerner, and L. J. Sansone. Experimental Behavior of Polymeric Geogrids in Pullout. *Journal of Geotechnical Division*, ASCE, Vol. 120, N. 4, 1994, pp. 661–677.
6. Holtz, R. D., and B. B. Broms. Wall Reinforced by Fabrics—Results of Model Test. *Proc., 1st International Conference on the Use of Fabrics in Geotecnics*, Paris, France, Vol. 1, 1977, pp. 113–117.

Publication of this paper sponsored by Committee on Geosynthetics.

One-Dimensional Compression Characteristics of Artificial Soils Composed of Multioriented Geosynthetic Elements

EVERT C. LAWTON, JAMES R. SCHUBACH, RICHARD T. SEELOS, AND NATHANIEL S. FOX

Results and conclusions are presented from laboratory one-dimensional primary and secondary compression tests on specimens of artificial soils composed of multioriented geosynthetic elements. The purpose of the tests was to determine the influence of the parameters on the one-dimensional compression characteristics of artificial soils composed of the following elements: properties of the polymeric constituent material and size and geometric shape of the elements. Values of one-dimensional primary and secondary compression parameters for these artificial soils are also compared with those for natural soils.

In previous work, the use of multioriented geosynthetic inclusions for reinforcing granular soils was investigated (1,2). In addition, the potential use of multioriented geosynthetic elements as lightweight, highly porous artificial soil was noted, and laboratory California bearing ratio (CBR) and permeability tests on specimens composed entirely of these elements were conducted. In the present study, the one-dimensional compression characteristics of artificial geosynthetic soils are examined, and typical results and conclusions from 40 constant-stress primary and secondary compression tests are presented. Each specimen tested was composed of discrete, multioriented, geosynthetic elements of a single prototype; overall, 20 prototypes manufactured in three different shapes from six polymeric materials were tested. The goals of this study were to determine the influence of the following parameters on the one-dimensional primary and secondary compression characteristics of artificial geosynthetic soils: properties of the polymeric constituent material and size and geometric shape of the elements.

BACKGROUND

Previous laboratory tests on artificial soil specimens composed of multioriented geosynthetic elements have indicated the following properties (2): dry densities of the specimens ranged from 2.39 to 3.53 kN/m³ (15.2 to 22.5 pcf), which represent a reduction of about 80 to 90 percent compared with typical compacted soils; coefficients of permeability varied from 0.14 to 0.17 cm/sec (0.28 to 0.34 ft/min), which are comparable to those for clean sands and clean sand-gravel mixtures; and CBR values ranged from 0.7 to 3.7. It was concluded from this limited laboratory testing program that lightweight, highly porous artificial soils could be produced from multioriented geosynthetic elements but that additional research

Evert C. Lawton, James R. Schubach, and Richard T. Seelos, Department of Civil Engineering, University of Utah, 3220 Merrill Engineering Bldg., Salt Lake City, Utah 84121; Nathaniel S. Fox, The Earthwater Corp., 769 Lake Dr., Lithonia, Ga. 30058.

was needed to establish the stress-strain-strength and degradation characteristics of these artificial soils.

CURRENT STUDY

Multioriented Elements

Twenty artificial soil prototypes were manufactured using six polymeric materials, six basic shapes, two sizes, and two stem shapes (Figure 1, Tables 1–3). All the prototypes consist of six stems extending radially from a central hub; 16 prototypes had enlarged heads on four stems, and four prototypes had no heads on any stems. All prototypes were cast in specially designed, two- or three-cavity injection molds. Where heads were cast on the end of the stems, manufacturing limitations permitted heads to be included on only four of the six stems. The multioriented elements are generally similar in shape and size to toy jacks and are therefore referred to as "jacks" for brevity.

The size of the headless prototypes is defined as the distance from the outer tips of either two corresponding heads (for prototypes with heads) or two corresponding stems (headless prototypes) measured along a longitudinal axis passing through the center of two stems in parallel (including the hub). The size of all prototypes with heads was 25.4 mm (1 in.). Headless prototypes were manufactured in both 19.1 and 25.4 mm (¾ and 1 in.) sizes.

The first shape (called "Original" herein) was used in previous studies (1,2) and is the most complex shape used in this investigation. The stems of the Original shape are square prisms, with the cross-sectional area of the four stems that support the heads being less than the headless stems; the heads are cubes and their cross-sectional area is greater than the cross-sectional area of the stems to which they are attached (Figure 1).

The stems of the other five basic shapes have the same nominal cross-sectional area (10.1 mm²) and vary according to the size of the elements (19.1 or 25.4 mm as previously defined). The shape of the stems and heads (square prism or cylinder) and the absence of stems or orientation of the heads relative to the two headless stems are described as follows (see Figure 1): (a) headless, (b) all four heads parallel to the headless stems (rocket shape), (c) all four heads perpendicular to the headless stems (pinwheel shape), and (d) two heads parallel and two heads perpendicular to the headless stems (up-down shape).

Selected properties of the six polymers from which the prototypes were made are given in Table 2. The materials are appropriate for injection molding manufacturing processes and were selected to

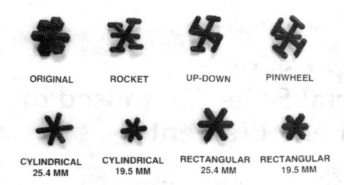

ORIGINAL ROCKET UP-DOWN PINWHEEL

CYLINDRICAL 25.4 MM CYLINDRICAL 19.5 MM RECTANGULAR 25.4 MM RECTANGULAR 19.5 MM

FIGURE 1 Geometric shapes of multioriented geosynthetic prototypes.

provide a wide range of strength and deformation characteristics. The strongest and stiffest material is the 20 percent glass-filled polypropylene (PPG) with a flexural modulus (E) of 4,479 MPa (650 ksi) and a tensile strength (F_t) of 55.1 MPa (8.0 ksi); the weakest and most flexible material is the low-density polyethylene (LDPE) with E = 282 MPa (41 ksi) and F_t = 8.3 MPa (1.2 ksi). Values of specific gravity range from a high of 1.24 for 40 percent mineral-reinforced polypropylene (PPM) to a low of 0.88 for copolymer polypropylene (PPPE).

Although the same mold was used to manufacture jacks of the same shape, the dimensions of the jacks made from different polymers varied somewhat from the nominal values. For example, the total volume per jack for the Original shape varied from 1,650 mm³

(0.101 in.³) for high-density polyethylene (HDPE) to 1,830 mm³ (0.112 in.³) for PPM.

Experimental Procedures

One-dimensional, constant-stress primary and secondary compression tests were performed on dry specimens of artificial jack soils confined within 152-mm (6-in.)-diameter steel molds. In the primary compression tests, loads were applied in increments to the maximum applied stress (σ_v) possible with the available equipment (300 kPa = 6.3 ksf). After the primary compression tests were complete, the maximum stress was maintained for 7 days to measure secondary compression.

Pilot tests conducted on specimens of the same prototype but varying heights showed that the height of the specimen affected the results somewhat owing to friction between the jacks along the exterior of the specimens and the interior surfaces of the molds. However, the general stress-strain characteristics and relative relationships for different prototypes were essentially independent of specimen height. As a compromise between manufacturing cost and ratio of specimen height to element size, a specimen height of 102 mm (4 in.) was selected for the main testing program. Thus, for specimens composed of 25.4 mm (1 in.) jacks, the ratio of specimen diameter to jack size was about 6:1 and the ratio of specimen height to jack size was about 4:1. For the 19.1 mm (¾ in.) jacks, the ratios were about 8:1 and 5.3:1.

To avoid undesirable gaps around the edges of the specimens, the specimens were made by placing jacks one by one into the mold until the final height was achieved. Thus, the relative densities of

TABLE 1 Shapes and Nominal Dimensions of Multioriented Inclusions

Shape	Size[a] (mm)	Stem Shape	Head Shape	Stems Length (mm)	Stems Width or Diameter (mm)	Heads Length (mm)	Heads Width or Diameter (mm)	Nominal Total Volume (mm³)
Original	25.4	Square Prism	Cubic	3.97	3.18	6.35	6.35	1,760
		Square Prism	None	11.11	4.76	None	None	
Headless Rectangular	25.4	Square	None	11.11	3.18	None	None	704
Headless Rectangular	19.1	Square	None	7.94	3.18	None	None	512
Headless Cylindrical	25.4	Cylinder	None	11.11	3.58	None	None	704
Headless Cylindrical	19.1	Cylinder	None	7.94	3.58	None	None	512
Rocket[b]	25.4	Cylinder	Cylinder	7.94	3.58	12.70	3.58	1,080
		Cylinder	None	11.11	3.58	None	None	
Pinwheel[c]	25.4	Cylinder	Cylinder	7.94	3.58	12.70	3.58	1,080
		Cylinder	None	11.11	3.58	None	None	
Up-Down[d]	25.4	Cylinder	Cylinder	7.94	3.58	12.70	3.58	1,080
		Cylinder	None	11.11	3.58	None	None	

[a]Distance from the outer tip of either two corresponding heads (for jacks with heads) or two corresponding stems (for headless jacks) along a longitudinal axis passing through the center of two stems in parallel (including the hub).

[b]All four heads are parallel to the headless stems.

[c]All four heads are perpendicular to the headless stems.

[d]Two heads are parallel and two heads are perpendicular to the headless stems.

TABLE 2   Selected Properties of Polymeric Materials

Property	Polymeric Material					
	Polypropylene 20% Glass Filled (PPG)	Polypropylene Mineral Filled (PPM)	Polypropylene Homopolymer (PP)	Polypropylene Copolymer (PPPE)	High Density Polyethylene (HDPE)	Low Density Polyethylene (LDPE)
Tensile strength (MPa)	55.1	31.7	34.8	21.0	22.0	8.3
Elongation at yield (%)	2	10	11	10	10	850
Flexural Modulus (MPa)	4,480	2,340	1,720	1,070	1,070	282
Notched Izod Impact (J/cm)	0.53	0.27	0.37	1.07	1.05	0.75
Deflection Temperature (°C) @ 455 kPa	138	132	140	145	130	88
Specific Gravity	1.04	1.24	0.89	0.88	0.95	0.92

All values supplied by manufacturers or distributors

TABLE 3   Basic Characteristics of Multioriented Inclusions

Prototype					
Shape	Polymer	Size (mm)	Mass per Jack (g)	Volume per Jack (mm^3)	Specific Gravity
Original	PPG	25.4	1.74	1,770	0.98
	PPM	25.4	2.13	1,830	1.16
	PP	25.4	1.58	1,780	0.89
	PPPE	25.4	1.53	1,740	0.88
	HDPE	25.4	1.54	1,650	0.93
	LDPE	25.4	1.57	1,750	0.90
Headless Rectangular	PPG	19.1	0.50	500	1.00
	PPG	25.4	0.68	700	0.97
	LDPE	19.1	0.44	490	0.90
	LDPE	25.4	0.61	690	0.88
Headless Cylindrical	PPG	19.1	0.44	460	0.95
	PPG	25.4	0.62	660	0.95
	LDPE	19.1	0.39	430	0.92
	LDPE	25.4	0.56	620	0.91
Rocket	PPG	25.4	1.01	1,050	0.96
Pinwheel	PPG	25.4	1.00	1,020	0.98
Up-Down	PPG	25.4	1.00	1,000	0.98
Rocket	LDPE	25.4	0.91	1,000	0.91
Pinwheel	LDPE	25.4	0.90	1,000	0.91
Up-Down	LDPE	25.4	0.91	1,000	0.91

the specimens were intermediate between a loose state representing a situation where the jacks would be dumped in place and a dense condition produced by vibratory compaction after being placed. Because there were differences in the dimensions and volume for jacks of the same shape but different material, there were inevitable differences in initial void ratio for nominally identical specimens. For example, for the Original shape jacks, the initial void ratios for the specimens made from the six different polymers varied from 2 to 2.8 (see Figure 2a).

Because the specimens were dry and the pore spaces were large, the pore air pressures generated during each loading increment were dissipated almost instantaneously. Therefore, the first reading

(15 sec) for each loading increment was assumed to represent the end of primary compression and the beginning of secondary compression. The specimen deformations were very large in some instances, so all deformation results are presented in terms of true strain instead of engineering strain, where true strain is given by the following equation:

$$\epsilon_v = \int_{H0}^{H} \frac{dH}{H} = \ln \frac{H}{H_0} \tag{1}$$

where H_0 is the original height of the specimen, and H is the height of the specimen at any time after loading.

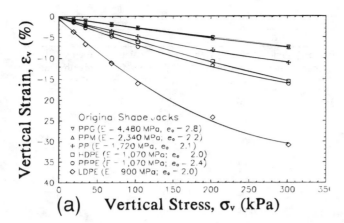

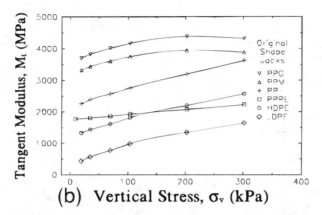

FIGURE 2   Effect of polymeric material type on one-dimensional primary compression of artificial jack soil: (a) strain versus stress, (b) tangent modulus versus stress.

Results From Primary Compression Tests

Effect of Polymer Type

The type of polymer used to manufacture the jacks has a significant effect on the relationship between applied vertical stress and primary compressive strain, as shown in Figure 2a for specimens of Original shape jacks. The flexural modulus of the plastic appears to be the most important factor; as flexural modulus increases, the compressibility of the artificial soil decreases. In one-dimensional compression, the decrease in volume results primary from two phenomena: (a) rearrangement of jacks from sliding along the stems and heads and (b) distortion of the jacks occurring primarily from bending of stems. Therefore, less bending of stems and less compression occur for stiffer plastics. As the applied load is increased, additional bending of stems occurs. At some point the bending of stems produces additional contact points between jacks, and the stiffness of the material increases. This is illustrated in Figure 2a by the flattening of the strain-stress curves with increasing stress. Note that the flattening of the curves is less for the stiffer plastics because less additional contacts are produced at lower values of strain. This stiffening effect is more clearly demonstrated in Figure 2b, where the results from Figure 2a are plotted in terms of one-dimensional tangent modulus (M_t) versus σ_v. For each specimen, M_t increases with increasing σ_v. Values for M_t for each specimen were obtained

by performing a polynomial least-squares regression on the strain-stress data with ϵ_v as the dependent variable and σ_v as the independent variable, differentiating the regression equation to obtain the slope of the strain-stress curve ($d\epsilon_v/d\sigma_v$), calculating the slope at the same levels of stress applied in the test, and inverting the slope to obtain the tangent modulus ($d\sigma_v/d\epsilon_v$).

Owing to the substantial amount of time required to manufacture the number of jacks required for a test—which ranged from about 350 to 1,600 jacks—the remaining prototypes were made from only the stiffest and strongest material, PPG, and the most flexible and weakest material, LDPE.

Influence of Stem Shape and Size

To determine the influence of the size and shape of the stems on the compressibility of artificial jack soils, headless jacks were manufactured using PPG and LDPE with two stem configurations (square prism and cylinder) and in two sizes (25.4 mm = 1 in. and 19.1 mm = ³/₄ in.). The nominal cross-sectional area of the stems for all four prototypes was the same (10.1 mm² = 0.0156 in.²). The lengths of the stems measured from the outer edge of the hub to the tip of the stems were 7.9 and 11.1 mm (⁵/₁₆ and ⁷/₁₆ in.) for the 19.1 mm (³/₄ in.) and 25.4 mm (1 in.) sizes, respectively. The results from primary compression tests conducted on artificial soil specimens made from these jacks (Figure 3) indicate that the length of the stem (for a given cross-sectional area) significantly affects the compressibility, but the cross-sectional shape of the stem does not. Both these trends are consistent with the assumption that one-dimensional compression of artificial jack soils occurs primarily from bending of the stems.

It can be shown that a solid square and a solid circle with the same area have the following ratio for their moments of inertia:

$$\frac{I_{sqr}}{I_{cir}} = \frac{\pi}{3} \tag{2}$$

For stems made from the same polymer (therefore same E) and with the same cross-sectional area, those with a square cross-sectional shape have a flexural rigidity (EI) about 5 percent greater than those with a circular cross-sectional shape. Thus, for specimens contain-

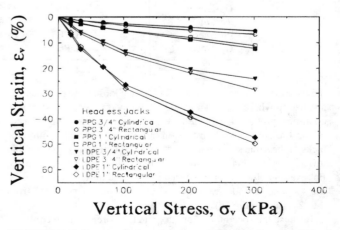

FIGURE 3   Influence of stem shape on strain-stress curves for headless jacks.

ing the same number of jacks, the load per jack, and therefore the load per stem, would be the same, and the specimens containing jacks with square prismatic stems should compress about 5 percent less than those with cylindrical stems. To compare the compression characteristics for the same size and same polymer but different cross-sectional shapes, the ratios of change in height for the comparable samples ($\Delta H_{sqr}/\Delta H_{cir}$) were calculated at each level of stress. For the 24 combinations of polymer, size, and stress level, $\Delta H_{sqr}/\Delta H_{cir}$ varies from 0.84 to 1.25, averaging 1.06, suggesting that the jacks with square stems are slightly more compressible than those with circular stems, which is opposite of the expected result. To account for the differences in number of jacks per specimen (N_J) caused by the slight differences in volume between the jacks with square stems and those with circular stems, and for random differences in placement, the deflections for each specimen were normalized by multiplying them by N_J. The calculated values of $(\Delta H \cdot N_J)_{sqr}/(\Delta H \cdot N_J)_{cir}$ range from 0.64 to 1.09, averaging 0.94. This average value suggests that the jacks with square stems are about 6 percent less compressible than those with circular stems, very close to the 5 percent predicted from Equation 3. Therefore, moment of inertia of the stems appears to affect the compression characteristics. Cross-sectional shape does not.

Substantial differences in stress-strain behavior are evident for the 25.4 mm (1 in.) and 19.1 mm (¾ in.) jacks made of the same polymer, as indicated in Figure 3. For the same polymer, the 25.4-mm (1-in.) jack specimens compressed an average of 82 percent more than the 19.1-mm (¾-in.) jack specimens. Two factors support this relationship: (a) the specimens containing larger jacks have fewer jacks per specimen (average $N_J = 1,409$) than those made of smaller jacks (average $N_J = 776$), and thus the load per jack is higher for the larger jacks; and (b) the stems of the larger jacks are longer and therefore more flexible. To provide additional insight into the effect of size on the one-dimensional compressibility of artificial jack soils, the following simplified theoretical relationship is developed.

Assumptions

- All deformations occur as a result of bending of stems.
- Each specimen contains an equivalent number of layers of jacks (N_L), with the same number of jacks in each layer. N_L is inversely proportional to the size of the jacks (S_J) and is not necessarily an integer.
- The vertical deflection of a specimen under an applied load is equal to the vertical deflection per jack times the number of layers ($\Delta H = \Delta H_J \cdot N_L$).
- Each jack in a specimen carries the same vertical force, $F_J = F_v/N_J$, where F_v is the total applied vertical force. Each jack compresses by the same amount (ΔH_J).
- N_J is inversely proportional to S_J.
- ΔH_J is inversely proportional to the flexural rigidity (EI) of the stems.
- ΔH_J is proportional to F_J.
- The vertical force carried by each stem in a jack is the same, and this force is applied at the tip of the stem. The deflection of each stem can be approximated by small deflection theory for a cantilever beam with a point load at the end of the beam. Therefore, ΔH_J is proportional to $L^3 \cos^2\theta$, where L is the length of the stem and θ is the angle of the longitudinal axis of the stem referenced to horizontal. The average value of θ is assumed to be the same for each specimen.

- The constant of proportionality for any specific relationship is independent of the other factors.
- The compression and extension of the stems caused by the longitudinal components of the applied forces is negligible compared with the deflections caused by bending of the stems. With these assumptions, the following ratio for deflections of two specimens supporting the same total vertical load (F_v) can be derived:

$$\frac{\Delta H_a}{\Delta H_b} = \frac{\left(\dfrac{L^3}{EI}\right)_a}{\left(\dfrac{L^3}{EI}\right)_b} \tag{3}$$

In Equation 3, the influence of size of jacks is evident only in terms of length of the stems; the effect of size on the number of jacks is offset by a proportional change in the number of layers.

For the situation where the size is varied but the polymer is the same [$(EI)_a = (EI)_b$], Equation 3 reduces to

$$\frac{\Delta H_a}{\Delta H_b} = \left(\frac{L_a}{L_b}\right)^3 \tag{4}$$

For the 25.4- and 19.1-mm jacks, the theoretical ratio of deflections based on Equation 4 becomes $\Delta H_{25.4}/\Delta H_{19.2} = 2.74$, compared with the actual average value of 1.82. The simplifications in this theory are numerous and a deviation from the theoretical ratio is expected. That the actual ratio is less than the theoretical ratio can be explained qualitatively in terms of major deviations from two of the assumptions. First, the deflections are quite large; hence large deflection theory is more appropriate. The deviation from small deflection theory increases as the factor PL^2/EI increases (3), where P is the bending force (perpendicular to the stems) and the other terms are as previously identified. Hence, the deviation from small deflection theory is greater for higher stresses, longer stems, and more flexible polymers, with the deflection perpendicular to the stem less than for small deflection theory and the deflection parallel to the stem greater than for small deflection theory. The deviation is greater for the perpendicular deflection than for the parallel deflection, so the net result for a stem oriented at 45 degrees to the horizontal is that the actual vertical deflection is less that predicted by small deflection theory, and the difference is greater for higher values of PL^2/EI. Using the average values for number of jacks and the lengths of stems, the ratio of PL^2/EI for the 25.4-mm (1-in.) jacks compared with the 19.1-mm (¾-in.) jacks is 3.6. Thus, the actual ratio of deflections should be smaller than the ratio predicted from small deflection theory (Equation 4), as is the case.

A second factor that tends to reduce $\Delta H_{25.4}/\Delta H_{19.2}$ is sliding of jacks relative to each other during loading owing to the smooth surfaces of the jacks. Because most of the sliding likely occurs along the stems of adjacent jacks, it would be reasonable to expect that the amount of sliding would be proportional to the length of the stems. Thus, the overall value of $\Delta H_{25.4}/\Delta H_{19.2}$ would be somewhere between $L_{25.4}/L_{19.1}$ for sliding and $(L_a/L_b)^3$ for bending.

Effect of Heads and Orientation of Heads

The effect of adding heads to the jacks and the effect of the orientation of those heads on the one-dimensional primary compression characteristics are illustrated in Figure 4 for jacks made from PPG

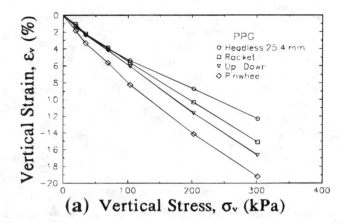

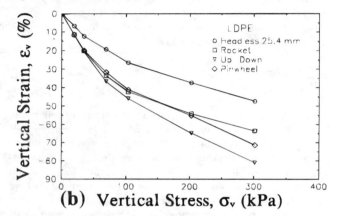

FIGURE 4 Comparison of strain-stress curves for jacks with and without heads: (*a*) PPG, (*b*) LDPE.

and LDPE. The jacks with heads in these tests essentially consisted of 25.4-mm (1-in.) cylindrical headless jacks with 3.2-mm (1/8-in.) heads on four stems oriented perpendicularly to the stems and extending 4.8 mm (3/16 in.) beyond the stems on each side; thus, the heads were 12.7 mm (1/2 in.) long from tip to tip. The volume of the cylindrical jacks with heads was about 50 percent more than the 25.4-mm (1-in.) cylindrical jacks without heads. The heads were oriented in three ways, as described previously, to obtain the three shapes (Pinwheel, Rocket, and Up-Down).

A comparison of the results in Figure 4 shows that for jacks made from PPG and LDPE, the specimens containing jacks with heads were more compressible than comparable specimens made of jacks without heads. Since the jacks with heads are essentially the same size as those without heads, N_L should be about the same for both. However, the jacks with heads require more space horizontally, so the number of jacks per specimen is much less for jacks with heads than those without heads; the average value of $N_{J-\text{headless}}/N_{J-\text{heads}}$ is 1.86 for PPG and 1.72 for LDPE. For jacks made from the same polymer (PPG or LDPE), the ratio of deflections for specimens of jacks with and without heads according to the theory described previously is as follows:

$$\frac{\Delta H_{\text{heads}}}{\Delta H_{\text{headless}}} = \frac{N_{J-\text{headless}}}{N_{J-\text{heads}}} \qquad (5)$$

The average actual value of $\Delta H_{\text{heads}}/\Delta H_{\text{headless}}$ for the three types of jacks with heads compared with the headless jacks for all six stress levels is 1.21 for PPG and 1.56 for LDPE. These values are less than those predicted from Equation 5, probably as a result of sliding among jacks during loading, as discussed previously. From the curves in Figure 4, it appears that the orientation of the head has some influence on the primary compression characteristics, with the Rocket shape seeming to be the least compressible, but with no discernible pattern as to which shape is most compressible. When the deflection results are corrected for the small variations in N_J, the average ratios of $\Delta H \cdot N_J$ for the other two orientations arbitrarily referenced to the Rocket shape are 1.04, 0.98, 1.02, and 1.42, suggesting that the differences in deflections probably result mainly from random variations in specimen preparation. Thus, the orientation of the heads seems to have little effect on the primary compression characteristics.

Artificial Jack Soils Versus Natural Soils

To assess the geotechnical characteristics of the artificial soils the load-deformation characteristics of the jacks were compared with a range of natural soils on the basis of Janbu's tangent modulus equation (*4*):

$$M_t = m \cdot \sigma_a \cdot \left(\frac{\sigma'}{\sigma_a} \right)^{1-a} \qquad (6)$$

where

m = modulus number,
a = stress exponent (number between 0 and 1), and
σ_a = reference stress = 1 atmosphere.

Values for m and a were determined for all artificial jack specimens tested in primary compression and are summarized in Table 4. Values of m give a general indication of the compressibility of the material; for the same value of a, lower values of m indicate greater compressibility. Values of a indicate the influence of stress level on M_t. For $a = 1$, M_t is independent of stress level, which is typical for many types of intact rock. A value of $a = 0$ indicates that M_t is linearly proportional to σ', which is characteristic of normally consolidated saturated clays. For many granular soils, a is about 0.5 (*5*).

Values of a and m for the artificial jack specimens as a function of porosity are compared with typical values for natural soils in Figure 5. The porosities of the artificial jack specimens are within the same range as for clays and peats. Both the type of polymer and geometric shape significantly affect the values of the tangent modulus parameters. The values of a for the artificial soils vary from 0.22 to 1. The lower values are for the LDPE jacks and are within the same range as for silts; the higher values are for the PPG jacks and are in the same range as for sands and moraines. The Original shape jacks—except for those made from LDPE—have a $\geq$ 0.76. Values of m for the artificial jack soils range from 4 to 65. The values for the LDPE jacks vary from 4 to 11 and are comparable to those for peats and soft clays. Values of m for PPG jacks and Original shape jacks (except LDPE) vary from 16 to 65, which compares with natural soils ranging from stiff clays to moderately dense silts. Note that the modulus number for all PPG specimens falls above the upper limit normally found for natural soils of the same porosity, indicating that for the same porosity the artificial jack soils are stiffer than natural soils.

TABLE 4 Values of Modulus Number and Stress Exponent for Multioriented Inclusions

Prototype Shape	Polymer	Size (mm)	Modulus Number m	Stress Exponent a
Original	PPG	25.4	41.9	0.94
	PPM	25.4	39.5	0.94
	PP	25.4	29.0	0.83
	PPPE	25.4	18.5	0.76
	HDPE	25.4	20.0	0.84
	LDPE	25.4	9.6	0.51
Headless Rectangular	PPG	19.1	43.7	0.65
	PPG	25.4	33.5	0.75
	LDPE	19.1	10.0	0.48
	LDPE	25.4	5.9	0.38
Headless Cylindrical	PPG	19.1	64.6	0.68
	PPG	25.4	22.9	0.73
	LDPE	19.1	11.0	0.22
	LDPE	25.4	6.2	0.35
Rocket	PPG	25.4	20.9	1.00
Pinwheel	PPG	25.4	15.9	0.78
Up-Down	PPG	25.4	18.8	1.00
Rocket	LDPE	25.4	5.2	0.49
Pinwheel	LDPE	25.4	3.9	0.44
Up-Down	LDPE	25.4	3.8	0.40

Results From Secondary Compression Tests

A series of secondary compression tests was conducted on specimens of jacks made after the conclusion of the primary compression tests. Typical results are shown in Figure 6 for the Original shape jacks. Approximate values for modified secondary compression index ($C_{\alpha\epsilon}$) were calculated for each test by determining a least-squares best-fit linear equation for $\epsilon_v = f[\log(t)]$ for each plot and calculating $C_{\alpha\epsilon}$ as the first derivative of the equation:

$$C_{\alpha\epsilon} = \frac{d\epsilon_v}{d[\log(t)]}$$ (7)

where t is the time after loading. The calculated values of $C_{\alpha\epsilon}$ are listed in Table 5 and vary from about 0.3 to about 2.8 percent. From these data, it is clear that the same relationships established for primary compression are also valid for secondary compression. In general, $C_{\alpha\epsilon}$ decreases (a) as M_t decreases, (b) as the size of the jacks increase, and (c) if heads are added to the jacks. No definite trends can be established for either the cross-sectional shape of the stems or the orientation of the heads. For saturated fine-grained natural soils, typical values of $C_{\alpha\epsilon}$ range from about 0.15 to 15 percent (6); hence, values of $C_{\alpha\epsilon}$ for artificial jacks soils are within the lower end of the saturated fine-grained soil range.

It was also desired to establish secondary to primary compression index ratios for the artificial soil specimens. To determine values for $C_{c\epsilon}$ at $\sigma'_v = 300$ kPa, the results from the primary compression tests were plotted in log $(\sigma'_v) - \epsilon_v$ space, as illustrated in Figure 6 for the Original shape jacks. Because the curves are not linear, a least-squares polynomial regression was performed on the data for each test, and tangent values for $C_{c\epsilon}$ were calculated by differentiating the polynomial and inserting $\sigma'_v = 300$ kPa into the equation for the first derivative:

$$C_{c\epsilon} = \frac{d\epsilon_v}{d[\log(\sigma'_v)]}$$ (8)

The calculated values of $C_{c\epsilon}$ varied from 0.062 to 0.84 (Table 5). Also shown in Table 5 are values of $C_{\alpha\epsilon}/C_{c\epsilon}$ for the artificial jack specimens, which varied from 0.027 to 0.11. Values $C_{\alpha\epsilon}/C_{c\epsilon}$ for the PPG specimens varied from 0.050 to 0.093 with an average of 0.071. For the LDPE specimens, $C_{\alpha\epsilon}/C_{c\epsilon}$ ranged from 0.027 to 0.089 with an average of 0.051, suggesting that $C_{\alpha\epsilon}/C_{c\epsilon}$ may be greater for jacks made from stiffer polymers than for those made from more compressible polymers. Note that the range in values of $C_{\alpha\epsilon}/C_{c\epsilon}$ for artificial jack soils is nearly the same as the range for natural soils (0.03 to 0.1) (7).

SUMMARY AND CONCLUSIONS

A laboratory experimental study consisting of one-dimensional primary and secondary compression tests was conducted to determine the one-dimensional compression characteristics of artificial soils consisting of multioriented geosynthetic elements (jacks). From the results of these tests, the influence of the following parameters on the one-dimensional compression characteristics of the artificial soils was assessed: geometric shape and size of the jacks and the

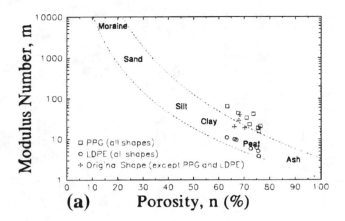

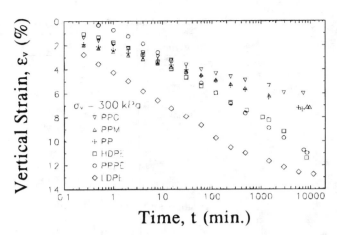

FIGURE 6 Secondary compression curves for original shape jacks.

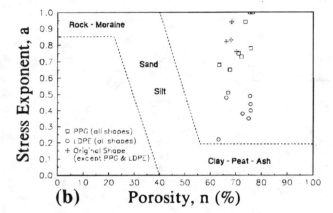

FIGURE 5 Values of Janbu's tangent modulus parameters for artificial jack soils as function of porosity: (*a*) modulus number, (*b*) stress exponent.

flexural properties of the polymer from which the jacks were made. The general results and conclusions determined from this experimental study are summarized as follows.

1. A substantial portion of the one-dimensional compression of artificial jack soils occurs from bending of the stems of the jacks about the central hub.

2. For a given geometric shape and size of the jacks, the flexural modulus of the polymer used in manufacturing the jacks has the greatest influence on the one-dimensional primary and secondary compression behavior of artificial jack soils. Primary and secondary compression are reduced for increased flexural modulus of the polymer.

TABLE 5   Values of $C_{\alpha\varepsilon}$, $C_{c\varepsilon}$, and $C_{\alpha\varepsilon}/C_{c\varepsilon}$ for Artificial Jack Soils

Prototype Shape	Polymer	Size (mm)	$C_{\alpha\varepsilon}$	$C_{c\varepsilon}$	$C_{\alpha\varepsilon}/C_{c\varepsilon}$
Original	PPG	25.4	0.011	0.12	0.092
	PPM	25.4	0.012	0.12	0.10
	PP	25.4	0.013	0.16	0.081
	PPPE	25.4	0.026	0.23	0.11
	HDPE	25.4	0.022	0.24	0.092
	LDPE	25.4	0.023	0.38	0.061
Headless Rectangular	PPG	19.1	0.0062	0.091	0.068
	PPG	25.4	0.013	0.14	0.093
	LDPE	19.1	0.019	0.35	0.054
	LDPE	25.4	0.027	0.57	0.047
Headless Cylindrical	PPG	19.1	0.0034	0.062	0.055
	PPG	25.4	0.0087	0.17	0.051
	LDPE	19.1	0.025	0.28	0.089
	LDPE	25.4	0.024	0.52	0.046
Rocket	PPG	25.4	0.016	0.24	0.067
Pinwheel	PPG	25.4	0.014	0.28	0.050
Up-Down	PPG	25.4	0.024	0.27	0.089
Rocket	LDPE	25.4	0.028	0.50	0.056
Pinwheel	LDPE	25.4	0.020	0.73	0.027
Up-Down	LDPE	25.4	0.026	0.84	0.031

3. For the same configuration and size of jacks, flexural rigidity of the stems significantly affects the primary compression behavior, whereas cross-sectional shape has little or no influence.

4. For the same basic geometric shape of jacks, artificial soil specimens made of larger jacks are more compressible than those composed of smaller jacks.

5. Specimens containing jacks with heads are more compressible than specimens made from similar jacks without heads.

6. A simplified theory based on small bending deflection theory was developed for comparisons of primary compression of artificial soils containing jacks with different characteristics. Using this theory, the qualitative trends described in Items 2 through 4 were predicted by the theory, but quantitative values differed somewhat from the actual values.

7. Owing to the high porosities of artificial jack soils, their one-dimensional compression behavior is comparable to natural soils ranging from peat through moderately dense silt.

ACKNOWLEDGMENT

The research described in this paper was funded by the Small Business Innovation Research program of the National Science Foundation via Phase II Grant No. DMI-9110387.

REFERENCES

1. Lawton, E. C., and N. S. Fox. Field Experiments on Soils Reinforced with Multioriented Geosynthetic Inclusions. In *Transportation Research Record 1369*, TRB, National Research Council, Washington, D.C., 1992, pp. 44–53.
2. Lawton, E. C., M. V. Khire, and N. S. Fox. Reinforcement of Soil by Multioriented Geosynthetic Inclusions. *Journal of Geotechnical Engineering*, ASCE, Vol. 119, No. 2, 1993, pp. 257–275.
3. Timoshenko, S. P., and J. M. Gere. *Mechanics of Materials*. Van Nostrand Co., New York, 1972, pp. 208–211.
4. Janbu, N. Soil Compressibility as Determined by Oedometer and Triaxial Tests. *Proc., European Conference on Soil Mechanics and Foundation Engineering*, Wiesbaden, Germany, Vol. 1, 1963, pp. 19–25.
5. *Canadian Foundation Engineering Manual*, 2nd ed., Meyerhof, G. G., and B. H. Fellenius (eds). Canadian Geotechnical Society, 1985.
6. Mesri, G. Coefficient of Secondary Compression. *Journal of the Soil Mechanics and Foundations Division*, Vol. 99, No. SM1, ASCE, pp. 123–137.
7. Mesri, G., and P. M. Godlewski. Time- and Stress-Compressibility Interrelationship. *Journal of the Geotechnical Engineering Division*, Vol. 103, No. GT5, ASCE, pp. 417–430.

Publication of this paper sponsored by Committee on Soil and Rock Properties.

Stress-Strain and Strength Behavior of Staple Fiber and Continuous Filament–Reinforced Sand

S. D. STAUFFER AND R. D. HOLTZ

Laboratory triaxial compression tests were performed to examine the stress-strain behavior and strength characteristics of sands reinforced with randomly distributed staple fibers and continuous filaments. A medium uniform sand and a medium moderately well-graded sand, both with the same D_{50}, were tested in consolidated-drained triaxial compression with volume changes measured. The reinforcement consisted of randomly distributed 100-mm-long staple fibers and continuous filaments of an untwisted, multifilament 100 percent polyester yarn of the same type used in Texsol construction. Reinforcement concentration was 0.2 percent by weight of sand, the same commonly used in the field. Results indicate that randomly distributed staple fibers and continuous filaments increase the compressive strength, axial and volumetric strain at failure, and postpeak strength loss of the composite compared with unreinforced sand behavior. Filament reinforcement was found to contribute significantly more to the increase in stress-strain behavior than staple fiber reinforcement. Finally, a reinforced well-graded subangular sand had a greater increase in the stress-strain characteristics than a reinforced uniformly graded subrounded sand.

Reinforced soil is a composite material in which the strength of the soil is enhanced by the addition of tensile reinforcement. The most notable example is of course reinforced earth, developed by Henri Vidal in France; steel strips are horizontally embedded in engineering fill to provide the reinforcement. Since the introduction of reinforced earth in 1966, alternative reinforcing materials—such as grids, sheets, and fibers made out of materials ranging from steel and other metals, plastics, and various synthetic polymers—have been developed and used extensively (1).

Another type of composite material developed in France, Texsol, is a mixture of randomly distributed continuous polyester filaments deposited simultaneously with sand using special equipment (2). The Texsol yarn provides tensile resistance to the sand, thereby greatly improving its strength and stability.

Composite materials consisting of randomly distributed staple fibers in granular soils have also been found to improve the strength properties of sand. The main advantages of randomly distributed staple fibers and continuous filaments when compared with horizontally oriented reinforcement are the absence of potential planes of weakness and some construction simplicity.

Previous studies have focused on the behavior of either staple fibers or continuous filaments, instead of comparing both. Wargo-Levine (3) looked at both staple fiber-reinforced and continuous filament–reinforced sand, specifically the influence of fiber and filament characteristics on the composite properties.

This paper describes an experimental program to determine the stress-strain behavior and strength characteristics of randomly distributed staple fiber (SF)–reinforced sand composites and randomly distributed continuous filament (CF)–reinforced sand composites. In addition, the influence of sand granulometry (i.e., gradation, particles size, and shape) on continuous filament composites was also investigated.

REVIEW OF PAST WORK

Composites With Staple Fibers

Triaxial tests were used to investigate the influence of various percentages of randomly oriented staple fibers on the properties of a granular soil (4). Although the presence of fibers decreased the density of the composite, the fibers increased the shear strength and failure strain of the sand tested (4).

Triaxial tests were also performed to examine the stress-strain and strength response of SF-reinforced sands (5). It was observed that (a) staple fibers increased the ultimate strength and stiffness of the composites, (b) shear strength increased linearly with increased fiber content until it reached an asymptotic upper limit of 2 percent by weight, and (c) the strength increase was more dependent on the surface friction of the fiber than the modulus (5).

A comprehensive experimental program was conducted to investigate the constitutive behavior of SF composites (6,7). It was found that the addition of randomly distributed staple fibers increased the shearing strength and stiffness, compared with unreinforced sand. An increase in the fiber aspect ratio, C_u and sand angularity, or a decrease in mean grain size increased the shear strength or both. Finally, a planar failure surface oriented at $45° + \phi/2$ was observed in the composite (6,7).

CONTINUOUS-FILAMENT COMPOSITES

Soil reinforced with continuous filaments (Texsol) was developed in France in the early 1980s. Some preliminary test results (8) showed great increases in the shear strength of sands with only 0.2 percent by weight of continuous filaments. Later a comprehensive series of triaxial tests to study the mechanical behavior of randomly distributed CF-reinforced sands was performed (9). It was found that (a) CF composites showed an increase in shear strength and strain at failure when compared with unreinforced soil, (b) the internal friction angle of the composite was always greater than that of

S. D. Stauffer, GeoEngineers, 8410 154th Avenue N.E., Redmond, Wash. 98052. R.D. Holtz, Department of Civil Engineering, University of Washington, FX-10, Seattle, Wash. 98195.

unreinforced soil, and (c) an increase in weight percentage of filament resulted in a greater value of pseudo or apparent cohesion.

The mechanical behavior of Texsol was studied (*10*) by performing triaxial compression tests, uniaxial compression tests, and direct shear tests. The results showed that the compressive strength of the CF composites increased in direct proportion to the weight percentage of the filament and decreased with filament linear density.

The stress-strain behavior of CF-reinforced sands using triaxial compression tests was examined (*11*). It was concluded that the continuous filaments increased the compressive strength, postpeak strength loss, strain at failure, and angle of internal friction of the sand.

Staple Fiber– and Continuous Filament–Reinforced Sand

In the only previous study of both staple fiber– and continuous filament–reinforced sand, the stress-strain behavior and strength characteristics of these composites under both static and dynamic loading conditions were investigated (*3*). It was found, among other things, that (a) CF reinforcement makes a greater contribution to the strength of the composite than does SF reinforcement; (b) the addition of either staple fibers or continuous filaments to the sand increased the compressive strength, strain at failure, and postpeak strength loss of the composite compared with unreinforced sand; (c) the failure surface of all randomly distributed fiber–reinforced sands was planar and had an orientation of $45 + \phi_r/2$; and (d) failure in both types of reinforcement appeared to be due to simultaneous slipping, stretching, and rupture of the fibers and filaments.

Volume changes were not measured and tests were performed with only one sand (*3*). Thus, it was appropriate to determine the effect of sand granulometry (i.e., gradation, mean grain size, and shape) on the stress-strain and strength characteristics of continuous filament–reinforced sands.

EXPERIMENTAL PROGRAM

Consolidated-drained triaxial compression tests were performed to determine the stress-strain and strength behavior of SF and CF sand composites under static loads. Specimens of two different sands at the same relative density were tested unreinforced, or reinforced with 0.2 percent by weight of either 100-mm staple fibers or continuous filaments.

Sands and Reinforcement

Two sands, both with the same mean grain size, were selected for this study. One was a well-graded subangular sand (Mortar sand) and the other was a uniformly graded, subrounded sand (Lonestar #3 sand). The grain size distribution curves of the sands are presented in Figure 1; other selected properties are in Table 1.

A 100 percent polyester multifilament yarn, provided by Societé d'Application du Texsol, Paris, France, was used in the testing pro-

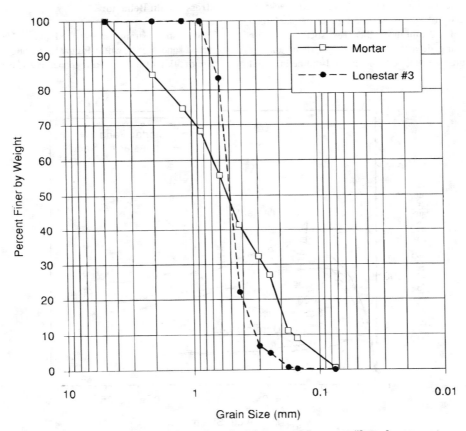

FIGURE 1   Grain size distribution curves for Mortar and Lonestar #3 sands.

TABLE 1 Properties of Sands Used in Experimental Program

Sand	D_{50} (mm)	C_u	G_s	e_{min}	e_{max}	D_r (%)	e	ρ_d (MN/m³)	Water Content (%)
Mortar	0.52	4.22	2.67	0.41	0.66	80	0.46	1.83	0.30
Lonestar #3	0.51	1.65	2.65	0.59	0.88	80	0.65	1.61	0.10

gram. The yarn consisted of 30 untwisted filaments with a linear density of 167 dtex and an elongation at rupture of 28 percent. A single piece of yarn over 100 m long was used for the CF composites. To make the staple fibers, the Texsol yarn was cut into staple fibers 100 mm long and in sufficient quantities to provide 0.2 percent by weight of the test specimen. This 0.2 percent value was selected because it is a common percentage used in the field (2).

Specimen Preparation and Testing

The sand and 0.2 percent reinforcement proportions were determined by considering the fibers and filaments as a solid in the voids-solid matrix. To avoid segregation of sand and reinforcement, a small initial water content (Table 1) was used. The components were mixed by hand until the reinforcement appeared to be randomly distributed throughout the sand. Individual test specimens were prepared inside a triaxial membrane supported by a split mold under vacuum. The moist tamping method (12) was used to achieve a uniform relative density of 80 percent throughout the specimen (71 mm in diameter and 152 mm in height). Test specimens were then saturated by flowing de-aired water through them and by applying a back pressure of 200 kPa. Consolidated-drained triaxial

compression tests under several different effective confining pressures were conducted with volume changes measured. The vertical load was applied by a material test system at a constant strain rate of 0.5 percent. Additional details of the experimental program can be found elsewhere (13).

TEST RESULTS

This section presents the results of the experimental testing program and compares these results with those of previous investigations on SF- and CF-reinforced sands.

Specific sand and reinforcement parameters were considered in terms of their influence on the (a) compressive strength, (b) axial strain at failure, (c) volumetric strain at failure, (d) stiffness, and (e) postpeak strength loss. In addition, the influence of the test variables on the Mohr-Coulomb strength parameters is also discussed.

Stress-Strain Behavior

Typical stress-strain curves for SF and CF composites and unreinforced sand are shown in Figures 2 through 4, for confining pressure of 50, 100, and 150 kPa, respectively. Figures 5 and 6 show the

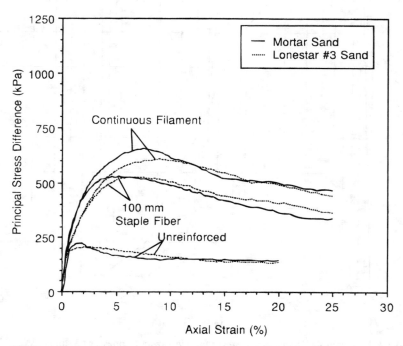

FIGURE 2 Representative stress-strain curves at 50 kPa confining pressure.

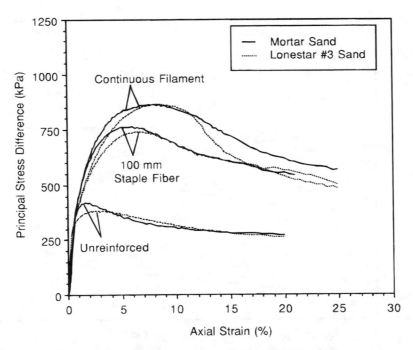

FIGURE 3   Representative stress-strain curves at 100 kPa confining pressure.

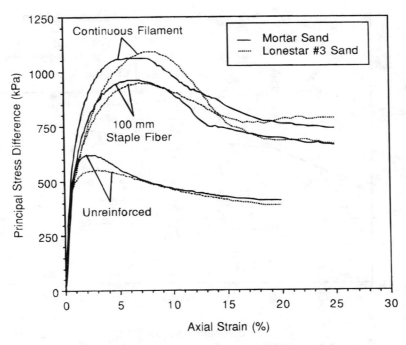

FIGURE 4   Representative stress-strain curves at 150 kPa confining pressure.

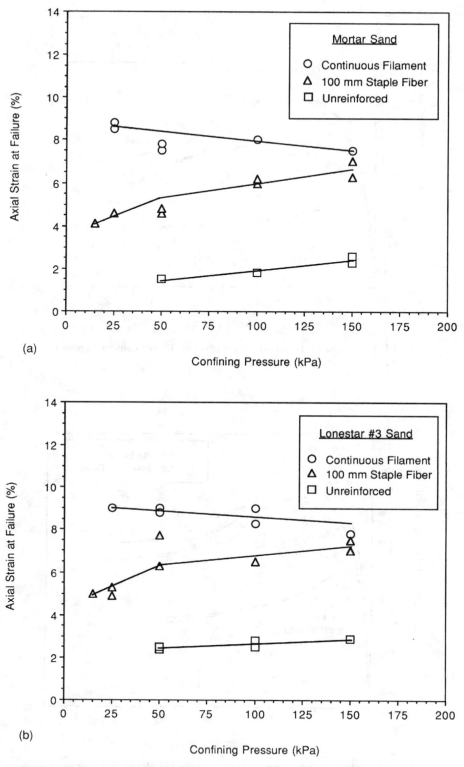

FIGURE 5 Axial strain at failure versus confining pressure for (*a*) Mortar sand and (*b*) Lonestar #3 sand.

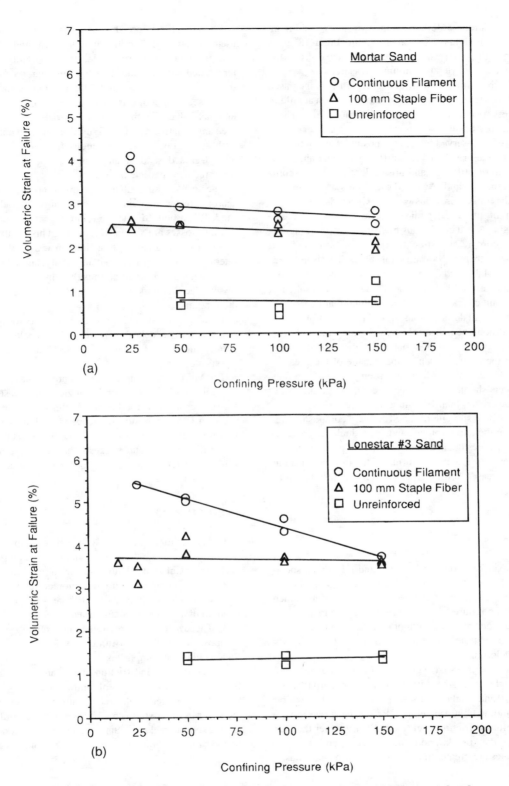

FIGURE 6 Volumetric strain at failure versus confining pressure for (*a*) Mortar sand and (*b*) Lonestar #3 sand.

axial and volumetric strains at failure for the two sands, and Figure 7 shows the secant modulus at 2 percent strain. Figure 8 gives the postpeak strength loss (residual stress subtracted from the maximum principal stress) versus confining pressure.

For both sands, the randomly distributed staple fibers increased the principal stress difference, axial and volumetric strain at failure, stiffness of the test specimens, and postpeak strength loss. These results agree with previous investigations.

Similar results were observed for CF composites. The continuous filaments increased the principal stress difference, axial and volumetric strain at failure, stiffness, and postpeak strength loss compared with unreinforced sand. Again, these results are in general agreement with the limited previous work on CF sands.

A comparison of the stress-strain behavior of SF- and CF-reinforced sands shows that continuous filaments contribute significantly more to the compressive strength, axial strain at failure, and postpeak strength loss than do 100-mm staple fibers. This is in agreement with works by Maher and Gray (6) and Gray and Maher (7) that reported that an increase in the length to diameter ratio resulted in improved stress-strain characteristics. It was found that fibers were placed in tension as small deformations took place in reinforced soil (14). This tension mobilized shearing stresses along the soil-reinforcement interface to large distances outside of the shearing plane. The likelihood of the reinforcement being outside of the shearing plane is greater in CF composites than in SF composites. Therefore, continuous filaments should be expected to contribute more to the stress-strain characteristics than 100-mm staple fibers.

Although the residual strength of the SF and CF composites was always greater than the unreinforced sand, it gradually approaches to a certain extent the residual strength of unreinforced sand. This appears to be caused by gradual rupturing of the reinforcement as the axial strains increase. Additional analysis of the results can be found elsewhere (13).

Strength Parameters

The results of triaxial tests on unreinforced and reinforced specimens are given in Figures 9 through 11. Figure 12 shows all failure envelopes. Friction angles and cohesion intercepts determined using Mohr-Coulomb failure criterion are summarized in Table 2.

Both reinforced sands revealed a minor cohesion intercept, probably the result of the straight line approximation of the slightly curved failure envelopes.

Both SF and CF specimens have greater cohesion intercepts and larger friction angles than their unreinforced counterparts. As shown in Figures 10 and 12, the sand-staple fiber composites exhibit bilinear Mohr-Coulomb failure envelopes. Below the "break" in the envelope, the reinforced friction angles are slightly larger than for sand alone. Above the "break," both the intercepts and the reinforced friction angles are greater than for the unreinforced sands.

Influence of Sand Granulometry

An increase in soil gradation (higher C_u) resulted in an overall higher compressive strength for both the SF and CF composites (Figures 2–4). This is in general agreement with previous work (6).

SF- and CF-reinforced composites have a higher axial and volumetric strain at failure than unreinforced sands (Figures 5 and 6). The percentage increase in axial strain is slightly greater for the better graded Mortar sand than for Lonestar #3 for both reinforcements. The opposite is true for the volumetric strain at failure.

The secant modulus is slightly greater for the well-graded (Mortar) sand than for the uniformly graded (Lonestar #3) sand at the same relative density (Figure 7). The percentage increases for both staple fibers and continuous filaments are also greater for a well-graded sand at all strains, probably the result of better reinforcement-particle interaction.

The postpeak strength loss for unreinforced well-graded (Mortar) sand is greater than unreinforced uniformly graded (Lonestar #3) sand. When both SF and CF reinforcement is added, the postpeak strength losses are similar (Figure 8). This result was unexpected, considering the greater interaction for well-graded sand. The better interaction, along with a high angularity, should have caused more reinforcement breakage and resulted in greater postpeak strength losses. See the work by Stauffer (13) for a detailed discussion of these points.

With respect to sand granulometry, the well-graded (Mortar) sand has a higher cohesion intercept but a smaller friction angle than a uniformly graded (Lonestar #3) sand (Table 2). This result was unexpected.

The Mortar sand staple fiber composites have a greater cohesion intercept and reinforced friction angle than Lonestar #3 composites above the break in the envelope. Below the break, Lonestar #3 shows a larger cohesion intercept value but a smaller reinforced friction angle. The existence of a larger cohesion intercept in the uniformly graded sand suggests that its strength envelope has a slightly greater curvature than does the well-graded sand.

CONCLUSIONS

1. The addition of randomly distributed staple fibers or continuous filaments to sand results in increased compressive strength, stiffness, axial and volumetric strain at failure, and postpeak strength loss compared with unreinforced sand.

2. Continuous filaments contribute significantly more to the strength of a composite than do staple fibers.

3. The triaxial compression failure envelopes for staple fiber composites are bilinear. Bilinear envelopes are not evident in continuous filament composites.

4. Reinforced sand friction angles are greater than for unreinforced sands for both types of reinforcement.

5. A reinforced well-graded subangular sand shows greater increases in stress-strain characteristics than a reinforced uniformly graded subrounded sand. Sand granulometry has no apparent effect on the Mohr-Coulomb parameters of reinforced sands.

ACKNOWLEDGMENTS

The first author gratefully acknowledges the financial support of the Valle Scholarship and Scandinavian Exchange Program at the University of Washington. We thank the Societé d'Application du Texsol, Paris, France, for supplying the reinforcement for this research.

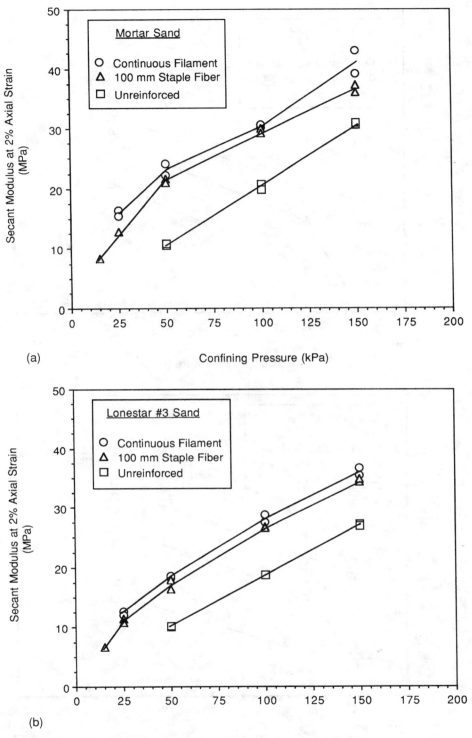

FIGURE 7 Initial tangent modulus versus confining pressure for (*a*) Mortar sand and (*b*) Lonestar #3 sand.

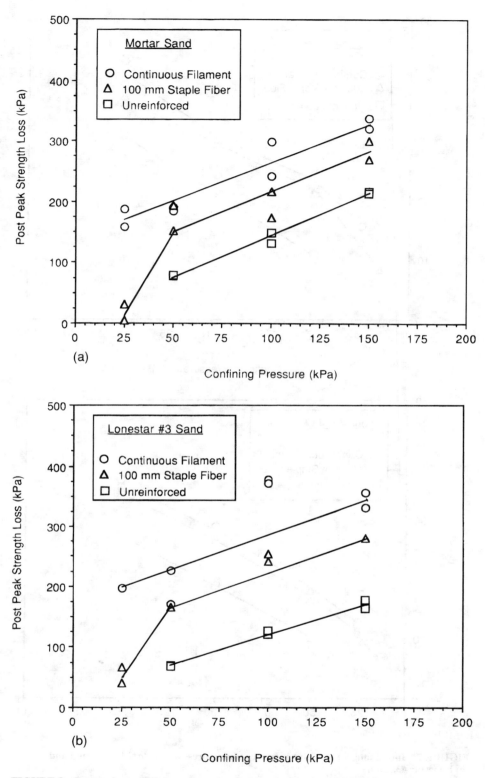

FIGURE 8 Postpeak strength loss versus confining pressure for (*a*) Mortar sand and (*b*) Lonestar #3 sand.

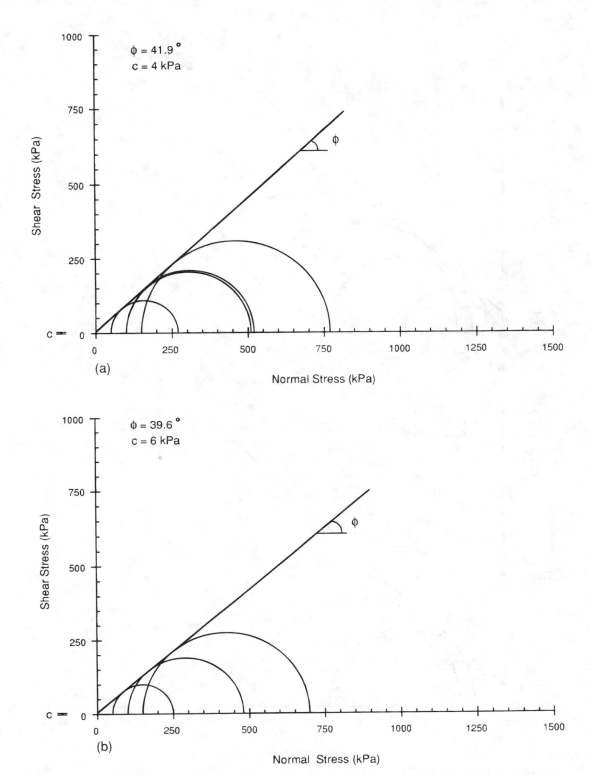

FIGURE 9   Mohr-Coulomb failure envelopes for unreinforced (*a*) Mortar sands and (*b*) Lonestar #3 sands.

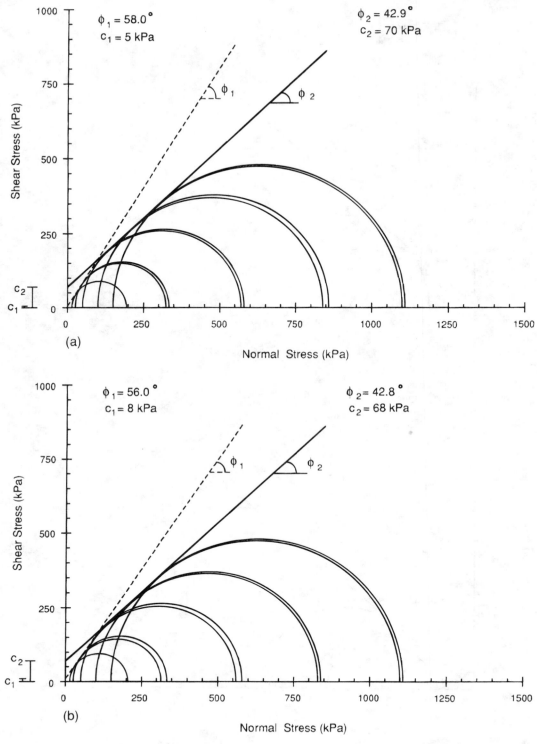

FIGURE 10 Mohr-Coulomb failure envelopes for (*a*) Mortar sands and (*b*) Lonestar #3 sands reinforced with 100-mm staple fibers.

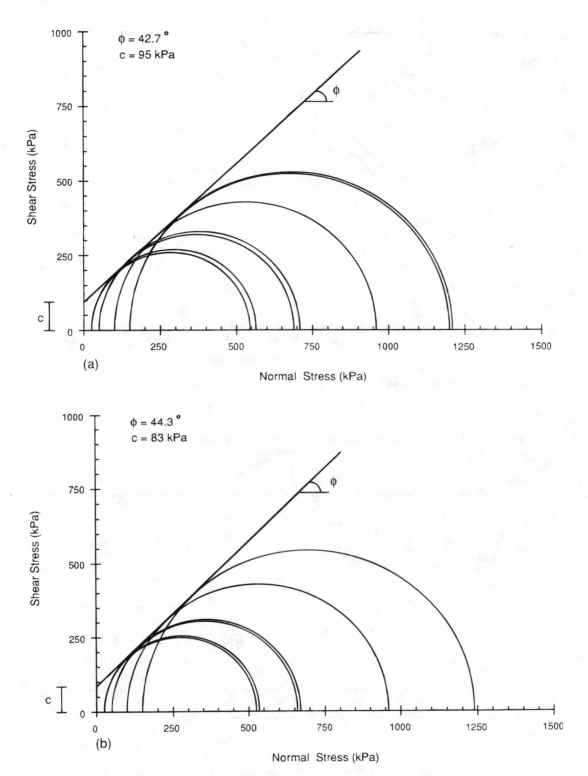

FIGURE 11 Mohr-Coulomb failure envelopes for (*a*) Mortar sands and (*b*) Lonestar #3 sands reinforced with continuous filaments.

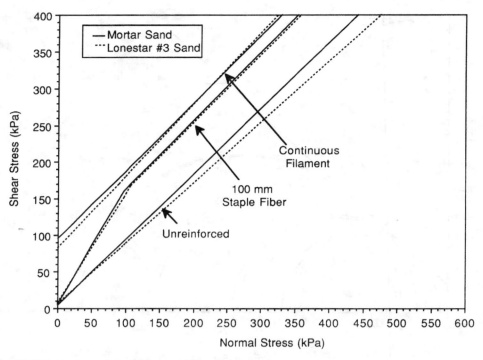

FIGURE 12   Mohr-Coulomb failure envelopes for all composite types.

TABLE 2   Strength Parameters from Mohr-Coulomb Failure Analysis

Reinforcement Type	Sand	Angle of Internal Friction (degrees)	Cohesion Intercept (kPa)
Unreinforced	Mortar	42	4
	Lonestar #3	40	6
100 mm Staple Fiber	Mortar	58	5
		43	70
	Lonestar #3	56	8
		43	68
Continuous Filament	Mortar	43	95
	Lonestar #3	44	83

REFERENCES

1. Hausmann, M. R. *Engineering Principles of Ground Modification.* McGraw-Hill, New York, 1990.
2. Leflaive, E. TEXSOL: Already More than 50 Successful Applications. *Proc., International Geotechnical Symposium on Theory and Practice of Earth Reinforcement,* Fukuoka, Japan, 1988, pp. 541–545.
3. Wargo-Levine, K. M. *The Effects of Static and Cyclic Loading on the Stress-Strain Behavior of Randomly Distributed Staple Fiber and Filament Reinforced Sand.* M.S. thesis, University of Washington, 1991.

4. Hoare, D. J. Laboratory Studies of Granular Soils Reinforced with Randomly Oriented Discrete Fibers. *Proc., International Conference on Soil Reinforcement,* Paris, France, Vol. I, 1979, pp. 47–52.
5. Gray, D. H., and T. Al-Refeai. Behavior of Fabric- Versus Fiber-Reinforced Sand. *Journal of Geotechnical Engineering,* ASCE, Vol. 112, No. 8, 1986, pp. 804–826.
6. Maher, M. H., and D. H. Gray. *Static and Dynamic Response of Sands Reinforced with Discrete, Randomly Distributed Fibers.* Final Technical Report to Air Force Office of Scientific Research, DRDA Project 024273, 1988.

7. Gray, D. H., and M. H. Maher. Admixture Stabilization of Sands with Random Fibers. *Proc., 12th International Conference on Soil Mechanics and Foundation Engineering*, Rio de Janeiro, Brazil, Vol. 2, 1989, pp. 1363–1366.

8. Leflaive, E. The Reinforcement of Granular Materials with Continuous Fibers. *Proc., 2nd International Conference on Geotextiles*, Las Vegas, Nev., Vol. 3, 1982, pp. 721–726.

9. Leflaive, E., and P. Liausu. The Reinforcement of Soils by Continuous Threads. *Proc., 3rd International Conference on Geotextiles*, Vienna, Austria, Vol. 4, 1982, pp. 1159–1162.

10. Khay, M., J. D. Gigan, and M. Ledelliou. Reinforcement with Continuous Thread: Technical Developments and Design Methods. *Proc., 4th International Conference on Geotextiles, Geomembranes, and Related Products*, the Hague, The Netherlands, Vol. 1, 1990, pp. 21–26.

11. Weerasinghe, R. B., and A. F. L. Hyde. Soil Reinforced with Continuous Filaments. *Proc., 4th International Conference on Geotextiles, Geomembranes, and Related Products*, The Hague, The Netherlands, Vol. 1, 1990.

12. Castro, G. Liquefaction of Sands. *Harvard Soil Mechanics Series No. 81*. Harvard University, Cambridge, Mass., 1969.

13. Stauffer, S. D. *The Stress-Strain Behavior of Sands Reinforced with Randomly Distributed Staple Fibers and Continuous Filaments*. M.S. thesis, University of Washington, 1992.

14. Shewbridge, S. E. *The Influence of Reinforcement Properties on the Strength and Deformation Characteristics of a Reinforced Sand*. PhD dissertation, University of California, Berkeley, 1987.

Publication of this paper sponsored by Committee on Soil and Rock Properties.

Influence of Geosynthetic Reinforcement on Granular Soils

I. ISMAIL AND G. P. RAYMOND

Model test results of strip footings on geosynthetic reinforced granular soil deposits are presented. The deposits consisted of a thin strong layer of granular material placed on a weaker granular material and a uniform single layer of the same weak granular material. The two-layer soil deposit was reinforced with a single layer of geosynthetic reinforcement. The uniform soil deposit was reinforced with one or two layers of geosynthetic reinforcement. The buried depth of the geosynthetic reinforcements was varied to determine the optimum position of placement. The optimum was based on maximizing the ultimate bearing capacity of the footing and reducing its settlement. The effect of repeated loading on the behavior of the geosynthetic reinforced granular deposits is also examined. The best method for improving the performance of weak granular soil deposits is concluded from the results.

Use of geosynthetic reinforcement grids (geogrids) in geotechnical structures, such as paved and unpaved roads, runways, and ballasted tracks, is increasing rapidly. Ballasted tracks, roads, and airfield pavements are examples of shallow foundations constructed using granular soils where the thicknesses of the layers are often relatively small compared with the width of the loaded area.

Ballasted tracks for large gantry cranes, built from granular material, are commonly subjected to very heavy loads. In trafficked areas, a thin top ballast layer is normally placed on top of a subballast layer. The top ballast generally consists of a crushed angular particle made from cobble sizes or quarried rock. The subballast is generally obtained from low-cost aggregates containing uncrushed rounded particles. Though it is potentially more economical to use uncrushed aggregate as the subballast, the subballast may cause a decrease in the stability and the track-holding capacity of the granular cover. Design problems for such construction will vary with their intended purposes. Interest might be with either a foundation failure under a concentrated load (as in the case of a gantry crane) or with trafficking problems due to rutting. One method of improving the load-bearing capacity and reducing the settlement of these tracks is to use a geosynthetic reinforcement. Relatively few studies are available relating to the optimum depth of geosynthetic reinforcement in granular soils. Studies by Dembicki et al. (1), Milligan and Love (2), Das (3), and Kinney (4) have evaluated the effects of placing geotextiles and geogrids at the interface of two different soils. However, there have been no investigations on the effects of placing the reinforcement at some other depth. This paper presents the results of such an investigation to determine the influence of the buried depth of the geosynthetic reinforcement in a uniform weak granular soil with a thin upper layer of stronger granular material or the geosynthetic reinforcement of the same uniform weak granular soil on the bearing capacity and settlement of a surface-supported footing. Model testing and the finite element method of analysis

were used in the investigation, although only the model testing is reported in detail here.

OBJECTIVE

The objective was to study, by means of model tests, the influence and comparison of geosynthetic reinforcement in granular soils using three different case deposits. The studies are illustrated schematically in Figure 1. The objectives of the individual case deposit studies were

- *Case 1*: to investigate the effect of a single layer of geosynthetic reinforcement on the bearing capacity and settlement of a footing placed on a thin layer of stronger granular material over a deep layer of weaker granular material,
- *Case 2*: to investigate the effect of two layers of geosynthetic reinforcement on the bearing capacity and settlement of a footing placed on a single layer of the weaker granular material, and
- *Case 3*: to investigate the effect of a single layer of geosynthetic reinforcement on the bearing capacity and settlement of a footing placed on a weaker single layer of the granular material.

FORMULATION OF EXPERIMENTAL PROGRAM

In this study, an experimental formulation was based on an approximate 10th scale for general rail track engineering practice. Ballast at 40-mm maximum size grading to 20-mm size was modeled by 4.8-mm (#4 sieve) grading to 2.4-mm (#8 sieve) aggregate. Ties at a length of 2 000 mm (typical for a gantry crane) interacting to form a continuous footing of that width were modeled by a plane strain (continuous) 200-mm-wide footing. A soil deposit through a rock cut could be as shallow as one-quarter the tie length (footing width). A deposit of approximately the footing width (200 mm) was selected. This ratio could be greater or less but testing in a work by Raymond et al. (5) has shown this to be a reasonable ratio. The microgrid used as reinforcement had a rib size of 0.3 mm or about one-tenth that of typical field geogrids. The minimum microgrid placement depth below the footing used in the study was 12.5 mm. This represents a ballast depth of 125 mm typically required to prevent geogrid damage from the tamper ties that are inserted below the ties to cause below rail-seat (rail-tie crossover) ballast compaction.

GEOSYNTHETIC PROPERTIES

The microgrid used was a biaxial-oriented polypropylene grid with approximately equal tensile strength in both directions. The main

Civil Engineering, Queen's University, Kingston, Ontario, K7L-3N6, Canada.

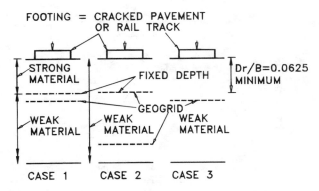

FIGURE 1   Schematic of cases studied.

properties were mass/unit area = 60 g\m³, ultimate tensile strength was approximately 48 kN/m and was independent of strain rate between 1 percent/minute to 0.001 percent/minute; strain at failure was between 10 to 12 percent, depending on testing speed. The faster testing rate resulting in the greater strain at failure. The stress-strain showed an approximately linear (slight curvature) response before failure.

TESTING EQUIPMENT

The layout of the testing equipment is shown in Figure 2. The tank used was 900 mm long, 200 mm wide, and 300 mm deep. The sides of the tank were made of herculite transparent glass with a very small coefficient of friction. The soils used were particles of a uniform 3.25-mm diameter rounded (weak material) and a similarly sized uniform graded crushed (strong material) ceramic Denstone made by Norton Chemical Processing Company. Both soils were repeatedly sized through a No. 4 (4.8-mm) sieve and retained on a No. 8 (2.4-mm) sieve to ensure a uniform grading free of broken smaller sizes. The particles had a specific gravity of 2.4. Their placement density was 1.51 and 1.40 g/m³ for the rounded and crushed particles, respectively. The soil was deposited in the test tank by dropping the particle through a uniform height of 300 mm. Dry drained triaxial tests on the rounded and crushed particles determined the internal angles of friction, ϕ', to be 34 degrees and 44 degrees respectively. A geosynthetic microgrid (geogrid) with an aperture size of 12.5 mm × 12.5 mm and a thickness of 0.3 mm was used in the tests. The geosynthetic reinforcement was cut to a length

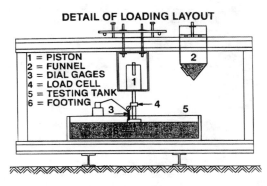

FIGURE 2   Schematic of test equipment.

and width 25.4 mm less than the length and width of the tank to prevent any friction between the geosynthetic reinforcement and the tank walls. The model footings were made from 19-mm-thick aluminum plate. They extended over the whole width of the tank. This simulated a plane-strain loading condition equivalent to track ties where the ballast arches between the ties, approximating a long footing of uniform width. Air pressure–activated loading pistons were used to load the footing. The loads were monitored by a load cell. Dial gauges, having a travel of 25 mm and sensitivity of 0.0025 mm, were placed near each of the four corners of the footing to monitor displacements. Four sets of thrust bearings, located on drilled seats in a rectangular plate, were used to ensure that the load always acted vertically on the footing.

EXPERIMENTAL STUDIES

Case 1

The first set of Case 1 tests consisted of statically loading a 200-mm-wide footing on a wide soil deposit in which the stronger upper layer was 12.5 mm thick and the lower weaker layer was 200 mm thick for a total depth of the two-layer deposit of 212.5 mm. Hereafter Ht and Hb will refer to the thickness of the top and bottom layers, respectively, and B will refer to the footing width. A single layer of geosynthetic reinforcement was placed at different depths below the surface, Dr, of 12.5, 25, 37.5, 50, 62.5, 75, and 125, and 175 mm, along with a test where no geosynthetic reinforcement was used. This gave ratios of geosynthetic reinforcement depth to footing width, Dr/B, of 0.0625, 0.125, 0.1875, 0.25, 0.3125, 0.5, 0.625, and 0.875 and the case of no geosynthetic reinforcement. The materials were then loaded statically to catastrophic failure resulting in movement to a purposely placed stop (settlement > 50 mm or $B/4$).

In the second set of tests, a single layer of geosynthetic reinforcement was placed at the same depths Dr as for the first set of tests, including the case of no geosynthetic reinforcement. This group of tests was then subjected to repeated loading. The repeated loadings were performed using a square wave at a frequency of 1 Hz, except for pauses at 1, 10, 10^2, 10^3, and 10^4 cycles. The pauses were made to apply a slow incremental applied load cycle of the same magnitude and lasted for about 1 hr. The pauses allowed the change in deformation modulus to be recorded. The moduli values are not presented here. Previous studies by Brown (6) showed little change in test observations after 10^4 loading cycles. A maximum average contact cyclic stress of 45 kPa was used. Tests (not presented here) established that for a single unreinforced soil layer, excessive settlement or failure resulted before 10^4 loading cycles when an average contact stress greater than 45 kPa was applied. After completing each repeated load test, the footing foundation was loaded to failure statically.

Case 2

The tests in Case 2 consisted of loading a 200-mm-wide footing on a wide 212.5-mm-deep deposit of the weaker granular soil. Thus, only one layer of granular material was used in this group of tests. Two layers of geosynthetic reinforcement were used. One layer of geosynthetic reinforcement was placed at a constant depth Dr of 12.5 mm ($Dr/B = 0.0625$). The second layer of geosynthetic reinforcement was placed at the same depths as for the tests of Case 1.

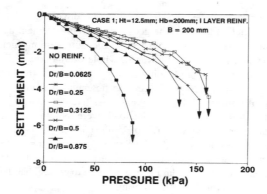

FIGURE 3 Load-settlement results before catastrophic failure for Case 1.

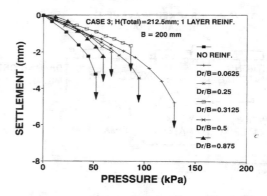

FIGURE 5 Load-settlement results before catastrophic failure for Case 3.

Also included was the case of no geosynthetic reinforcement. Similar to Case 1, two sets of tests—static and repeated load—were conducted, and at the end of the repeated load testing the footing foundation was loaded to failure statically.

Case 3

The tests in Case 3 were identical to the Case 2 tests in all respects except that the upper layer of geosynthetic reinforcement was illuminated. The single layer of geosynthetic reinforcement was thus varied in depth from an increasing ratio Dr/B of 0.0625 to 0.875.

STATIC TEST RESULTS

The static load was applied in small increments. Each increment was applied for 60 sec. The settlement was read after 40 sec. The load-settlement observations before catastrophic failure for the static set of tests performed in Cases 1, 2, and 3 are presented in Figures 3, 4, and 5, respectively. At low pressure levels, the settlement for all tests increased at an approximately constant rate. As failure was approached, the incremental rate of settlement increased until catastrophic failure occurred. Note that the subsequent load increment after failure, for every test, caused the maximum permitted movement of the loading piston. This was set to allow a footing settlement of at least 50 mm (i.e., > $B/4$). Herein, the ultimate bearing

capacity (UBC calculated as the average intensity of loading, q_u) used is the last stable load placed on the footing. It may be seen from the figures that the settlement patterns for tests in all cases were similar, and that the geosynthetic reinforcement had the effect of increasing the UBC (q_u) of the footing and decreasing the settlement at any given load. The general trend was for the higher UBC (q_u) to be associated with the stiffer settlement responses, although this was not true for every test result. Although the effect of reinforcement locations on both the UBC (q_u) and settlement is variable, significant improvement may be seen when $Dr/B < 0.5$ for Cases 1 and 2 and < 0.3 for Case 3. When the geosynthetic reinforcement is placed to give the most benefit (optimum depth), the UBC (q_u) was approximately doubled (or greater) and the settlement at the same load was reduced by approximately 50 percent or more.

The results for the UBC (q_u) versus Dr/B for Cases 1, 2, and 3 are plotted in Figure 6. All were tested on the same overall depth of granular material (i.e., 212.5 mm). The results of the tests on the uniform soil deposit with a single layer of reinforcement (Case 3) follow the same trend as previously reported (7–11,5). In this case, the closer the geosynthetic reinforcement is located to the footing base, the more effective the soil geosynthetic reinforcement. This does not occur for the two layered soil deposit of Case 1 or the uniform soil with two layers of geosynthetic reinforcement of Case 2. It may be seen from Figure 6 that the UBC (q_u) of the reinforced two-layer deposit (Case 1) and the uniform soil with two layers of geosynthetic reinforcement (Case 2) are very much governed by the depth of the geosynthetic reinforcement.

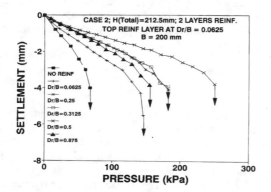

FIGURE 4 Load-settlement results before catastrophic failure for Case 2.

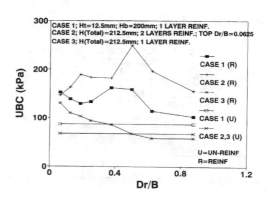

FIGURE 6 Variation of UBC (q_u) with Dr/B for Cases 1, 2, and 3.

In the case of the two-layer soil deposit (Case 1) when the geosynthetic reinforcement is placed very near the footing base, a high bearing capacity is first observed. As the depth of the geosynthetic reinforcement increased, the UBC (q_u) first decreased until the geosynthetic reinforcement depth to footing width ratio, Dr/B, equaled 0.1875. As the ratios of Dr/B then increased, the UBC (q_u) also increased until a maximum was observed at a ratio of Dr/B between 0.3 to 0.5, after which the UBC (q_u) decreased as Dr/B continued to increase. It is to be expected that had the geosynthetic reinforcement depth continued to be increased, a depth at which there would be a negligible effect from the introduction of the geosynthetic reinforcement could be identified. Indeed, it has been shown (9) that at a depth between $Dr/B = 1$ to 2, in a uniform soil deposit of depth to footing width ratio $H/B = 3$, the geosynthetic reinforcement had a negative effect [i.e., the UBC (q_u) decreased below that of the UBC (q_u) of an unreinforced deposit].

When two layers of geosynthetic reinforcement (Case 2) were introduced into the uniform soil deposit, the plot of the UBC (q_u) with Dr/B obtained gave trends similar to the Case 1 where a thin strong layer is placed on top of a deeper weaker layer. The UBC (q_u) increased as the depth to footing width ratio, Dr/B, of the second layer of the geosynthetic reinforcement increased reaching a maximum at a $Dr/B = 0.5$. The UBC (q_u) then decreased for values of Dr/B of the second deeper reinforcement greater than 0.5. During the experimental testing of Case 2, the geosynthetic reinforcement that was initially placed at values of $Dr/B \geq 0.5$ failed by breaking into two pieces below the center line of the footing. In view of this, the test was repeated several times using a number of stronger geogrids. Within experimental accuracy, all the tests using the same configuration gave the same test results. Herein, only the averages of the tests in which the geosynthetic reinforcement remained intact are reported.

The values of UBC (q_u) recorded at all Dr/B values for Case 2 are the highest values recorded in all the three cases investigated. This shows the advantages of using two layers of geosynthetic reinforcement. It must also be remembered that for Case 2, where two layers of geosynthetic reinforcement was used in a weaker soil, the UBC (q_u) was higher than the two-layer deposit with a thin stronger soil layer in the upper zone (Case 1). In an extension of the research (not presented here), it was observed that so long as the lower geosynthetic reinforcement is $\geq B$ and the upper geosynthetic reinforcement is $\geq 1.5\,B$, the same high UBC (q_u) are obtained (both geosynthetic reinforcements being centered below the footing). In fact, both these lengths may be reduced by a length of $0.5\,B$ each, and the geosynthetic reinforcement was observed to have some beneficial value. This means that beneficial effects may be achieved from small widths of geosynthetic reinforcement. Thus the cost of using two layers of geosynthetic reinforcement should not be a factor preventing adoption of this procedure.

REPEATED LOADING TEST RESULTS

Plots of the settlement versus logarithm of number of loading cycles for the footing on a two-layer granular deposit with geosynthetic reinforcement at various depths (Case 1) are presented in Figure 7. The curves characterizing the settlements all trend in the same nonlinear pattern. These plots show that the cumulative plastic settlements observed, at the same number of load cycles, decreased when the geosynthetic reinforcement was added. The plastic settlement is defined herein as the remaining settlement after the removal of the load.

Figure 8 shows, for all three cases, the variation of the cumulative plastic settlement under the maximum number of load applica-

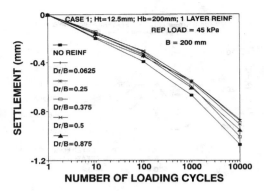

FIGURE 7 Typical plot of settlement with number of loading cycles.

tions applied prior to the static load testing to catastrophic failure. Generally the optimum geosynthetic reinforcement position is observed to be between values of $Dr/B = 0.3$ and 0.5. If the geosynthetic reinforcement is placed outside this range, the repeated load settlements are higher. Again, the uniform material with two layers of geosynthetic reinforcement (Case 2) is superior to the other two cases. Generally, the position of geosynthetic reinforcement was less significant in the two-layer granular deposit (Case 1) than it was in the uniform granular deposits (Cases 2 and 3).

Figure 9 shows for all three cases the UBC (q_u) obtained from the static load testing at the end of the 10^4 cycles of repeated loading and the values without repeated loading against the Dr/B ratio. For the repeated loading tests the UBC (q_u) was obtained by loading the soil deposits to failure statically after completing 10^4 cycles. The results show that repeated loading increased the UBC (q_u) above that obtained when no cyclic loading was applied for all deposits tested having the same dimensions and arrangement of geosynthetic reinforcement.

Note that, as state earlier, during the Case 2 static testing when the geosynthetic reinforcement was at a depth ratio of $Dr/B \geq 0.5$ the initial geosynthetic reinforcement used failed by breaking into two pieces directly below the center line of the footing. These tests were repeated using a stronger reinforcement, and only the static failure tests where the reinforcement remained intact are reported in Figure 9. The tests that were duplicated with a stronger geosynthetic reinforcement gave, within experimental accuracy, identical repeated loading results [i.e., the strength of the reinforcements (unfailed) substituted had no measurable effect on the repeated loading portion of the results].

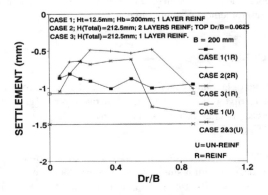

FIGURE 8 Variation of cumulative plastic settlement with Dr/B for Cases 1, 2, and 3.

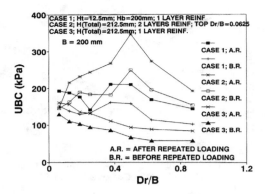

**FIGURE 9   Comparison of UBC (q_u) with Dr/B
for pre-cyclic (B.R.) and post-cyclic (A.R.) tests.**

NUMERICAL ANALYSIS

To analyze the results theoretically the finite element method (FEM) of analysis was used. Details of the program have been presented elsewhere (*11*). The FEM mesh was given the same dimensions as the experimental equipment. The tank ends restricted the horizontal displacements, and the tank bottom restricted the vertical displacements. The end walls and base were modeled as smooth (i.e., frictionless and nonadhesive). Eight node quadrilateral elements and an extended hyperbolic elastoplastic model with Mohr-Coulomb's failure criterion were used to model the soil. Beam elements with a high moment of inertia and a high lateral stiffness were used to model the rigid body motion of the footing. Three node bar elements were used to model the geosynthetic reinforcement. Six-node interface elements were used to model the friction between the soil and the footing, and the soil and the geosynthetic reinforcement. The angle of friction between the soil and the footing was taken as two-thirds of the angle of friction of the soil.

The FEM load-settlement curves failed to give catastrophic failures as recorded in the static tests; the results are therefore not given here. The results, however, showed the same trends as the static tests insofar as they predicted the best depth for placement of the reinforcement. For the uniform soil with two layers of geosynthetic reinforcement (Case 2), this occurred at a ratio Dr/B of approximately 0.3. The results also showed that the uniform soil deposit with two layers of reinforcement gave the best reinforcement benefit. This suggests promise for further work in refinement of finite element modeling for estimating the optimum placement of geosynthetic reinforcement configurations.

CONCLUSIONS

A number of laboratory model tests and FEM analyses for the UBC (q_u) of a strip footing were performed on two layers of geosynthetic reinforced granular material. Conclusions from the observations are as follows.

1. The results for Case 1 showed that, where a two-layer granular soil deposit with a single layer of geosynthetic reinforcement was tested, the UBC (q_u) was highest and the settlements for the same load lowest when the ratio of Dr/B was in the range of 0.3 to 0.5 (Figure 3).

2. Similar to Conclusion 1 for Case 1, the results for Case 2 showed that, where a uniform granular soil deposit with two layers of geosynthetic reinforcement was tested, the UBC (q_u) was highest and the settlements for the same load lowest when the ratio of Dr/B was in the range of 0.3 to 0.5 (Figure 4).

3. In contrast to Conclusions 1 and 2, the results for Case 3 showed that, where a uniform granular soil deposit with a single layer of reinforcement was tested, the UBC (q_u) decreased and the settlements for the same load increased as the geosynthetic reinforcement depth increased (Figure 5).

4. When the UBC (q_u) for the double geosynthetic reinforced uniform granular deposit (Case 2) was compared with the UBC (q_u) for either the singly reinforced two-layer granular soil deposit (Case 1) or the UBC (q_u) of the singly geosynthetic reinforced uniform granular deposit (Case 3), the UBC (q_u) for Case 2 is always higher for the same (lower) geosynthetic reinforcement positions (Figure 6).

5. When the UBC (q_u) for the single geosynthetic reinforced two-layer granular deposits (Case 1) is compared with the UBC (q_u) for the single geosynthetic reinforced uniform granular deposit (Case 3), the UBC (q_u) for Case 1 is always higher for the same geosynthetic reinforcement positions (Figure 6).

6. For any given reinforcement configuration, the static UBC (q_u) observed at the end of 10^4 cycles of repeated loading was greater than the static UBC (q_u) values observed when the load was not cycled before static loading (Figure 9).

7. A single layer of geosynthetic reinforcement positioned close to the footing base (at a ratio of $Dr/B = 0.0625$) in a uniform granular soil (Case 3) increased the UBC (q_u) and decreased the settlements at the same loads over the case of unreinforced soil (Figure 5). The reinforcement at shallow depths gave benefit trends similar to the effect of placing a thin stronger unreinforced granular layer used in this study on the weaker unreinforced granular deposit (shallow reinforced results for Case 3 in Figure 5 compared with unreinforced results for Case 1 in Figure 3).

8. A geosynthetic reinforcement at a depth ratio of $Dr/B = 0.3$ to 0.5 increased considerably the UBC (q_u) of (a) a two-layer soil having a thin upper stronger soil layer (Case 1) or (b) a uniform soil with an upper reinforcement layer (Case 2). Similarly the settlements were reduced.

9. Optimumly placed reinforcement reduced the cumulative plastic settlements caused by repeated loadings (Figure 8).

ACKNOWLEDGMENTS

The financial support provided by the Natural Scientific and Engineering Research Council of Canada in the form of a grant awarded to G. P. Raymond and the Canadian Commonwealth Scholarship and Fellowship Plan is gratefully acknowledged. The experimental tests were performed in the laboratories of Queen's University at Kingston, Ontario.

REFERENCES

1. Dembicki, E., P. Jermolowicz, and A. Niemunis. Bearing Capacity of Stripfoundation on Soft Soil Reinforced by Geotextile. *Proc., 3rd. International Conference Geot.*, 1986, Vol. 1, pp. 205–209.
2. Milligan, G. W. E., and J. P. Love. Model Testing of Geogrids under an Aggregate Layer on Soft Ground. *Proc., Sym. Polymer Grid Reinforcement*, Thomas Telford, 1984, pp. 128–138.

3. Das, B. M. Foundation on Sand Underlain by Soft Clay with Geotextile at Sand-Clay Interface. *Proc., Geosynthetics '89 Conference*, 1989, Vol. 1 pp. 203–214.

4. Kinney, T. Small Scale Load Tests on a Soil Geotextile Aggregate System. *Proc., 2nd. International Conference Geot.*, 1982, Vol. 2, pp. 405–409.

5. Raymond, G. P., M. S. A. Abdel-Baki, R. G. Karpurapu, and R. J. Bathurst. Reinforced Granular Soil for Track Support. *Grouting Soil Improvement and Geosynthetics*, Geotechnical Special Publication 30, ASCE, Vol. 2, 1992, pp. 1104–1115.

6. Brown, S. F. Repeated Load Testing of a Granular Material. *Journal of Geotechnical Engineering Division*, ACSE, Vol. 100, No. GT7, 1974, pp. 825–841.

7. Akinmusuru, J. O., and J. A. Akinbolade. Stability of Loaded Footings on Reinforced Soil. *Journal of Geot. Engng. Div.*, ASCE, Vol. 107, No. GT6, 1981, pp. 819–827.

8. Guido, V. A., D. K. Chang, and M. A. Sweeney. Comparison of Geogrids and Geotextiles Reinforced Earth Slabs. *Canadian Geotechnical Journal*, Vol. 23, 1986, pp. 435–440.

9. Raymond, G. P. Reinforced Sand Behavior Overlying Compressible Subgrades. *Journal of Geotech. Engng. Div.*, ASCE, Vol. 118, No. GT11, 1992, pp. 1663–1680.

10. Abdel-Baki, M. S., G. P. Raymond, and P. Johnson. Improvement of the Bearing Capacity of Footings by a Single Layer of Reinforcement. *Proc., Geosynthetics '93 Conference*, Vol. 1, 1993, pp. 407–416.

11. Abdel-Baki, M. S., and G. P. Raymond. Reduction of Settlements using Soil Geosynthetic Reinforcement. *Vertical and Horizontal Deformations of Foundations and Embankments*, ASCE, Geotechnical Special Publication 40, Vol. 1, 1994, pp. 525–537.

Publication of this paper sponsored by Committee on Soil and Rock Properties.

Limit Condition for Fiber-Reinforced Granular Soils

RADOSLAW L. MICHALOWSKI AND AIGEN ZHAO

Approximate methods for analysis and synthesis of earth structures using such reinforcement as geotextiles or geogrids have been used successfully for more than two decades. Soil reinforcement with short fibers and continuous synthetic filaments has been tried in the past decade. Although the latter has been used with success in construction practice, no techniques for design with such materials exist. This is mainly the result of poor understanding of the fiber-matrix (filament-matrix) interaction and, consequently, lack of appropriate models capable of describing the stress-strain behavior and failure of such composites. An attempt at describing the failure criterion of fiber-reinforced sand is presented. The properties of the constituents are approximated by their standard characteristics: the Mohr-Coulomb failure function for the granular matrix and the Tresca criterion for fibers. The failure of the composite is considered to be because of the collapse of the matrix associated with intact fibers or with fibers failing in slip or the tension mode. An energy-based homogenization technique is used. Results of some preliminary laboratory experiments are also presented. Application of the derived failure criterion is shown in an example of a slope limit load problem.

Although considerable progress in design methods with traditional reinforcement (geotextiles, geogrids, etc.) has been made in the past decade, no techniques for design with fiber-reinforced or continuous filament-reinforced soil exist. This is predominantly because of poor understanding of the behavior of such composites, even though successful applications are known (*1*).

This paper focuses on a failure criterion for fiber-reinforced soils using a plasticity-based technique. A limit surface is found in the macroscopic stress space, which describes stress states associated with failure of a fibrous composite with a granular matrix.

Elastic and elastoplastic behavior of composite materials has been described using various methods of homogenization (averaging), ranging from self-consistent schemes (*2–4*) to the finite element approach (*5*). An excellent survey of techniques used for analysis of composite materials was presented in a work by Hashin (*6*); since then, significant interest in fiber composites has been maintained.

Little attention has been paid to composites with granular or low-cementitious matrices. Failure criteria of such materials must be known to evaluate the stability of structures such as reinforced soil slopes. Existing literature includes only a handful of papers with attempts at describing theoretically the behavior of reinforced soil and, particularly, fiber-reinforced or continuous filament-reinforced granular composites (*7–12*). This paper presents a continuation of previous efforts toward constructing a consistent model for predicting failure of fiber-reinforced soils.

Experimental results from tests on specimens of fiber-reinforced soils are available (*13–17*). Development of a strength criterion based on considering behavior of fibers intersecting a strain localization zone (shear band) dominates among the theoretical efforts triggered by experimental observations. Strain localization should be avoided in mathematical homogenization schemes or when testing for material properties, because the boundary displacements are no longer representative of the strain of the entire material within a specimen. This does not contradict the usefulness of the direct shear test where parameters for a well-defined model are tested. A common interpretation of results from tests on fiber-reinforced samples is a piecewise linear failure condition, with the first range (at low confining stresses) relating to fiber slip and the second one associated with fiber yielding. Whereas such two clearly different ranges of behavior can be conjectured intuitively, a sharp transition, often suggested in the literature, appears to be the outcome of a loose interpretation of the test results.

This paper is intended to introduce the concept of homogenization and macroscopic description to fiber-reinforced soils. It presents the preliminary results, theoretical and experimental, but, for brevity, the details of theoretical derivation and experimental techniques are omitted here.

Fundamentals of the homogenization technique used will be presented in the next section, followed by a brief description of the failure criterion for fiber-reinforced soil. Next, some experimental results from triaxial tests on specimens of sand reinforced with steel and polyamide fibers will be presented. The paper is concluded with an example of the application of the derived failure criterion and some final remarks.

MACROSCOPIC LIMIT STRESS

The term "macroscopic stress" is used here to represent the average stress in the composite. Because the failure criterion is of interest, the macroscopic stress at the limit state is investigated. The homogenization technique used here is based on consideration of the energy dissipation during plastic deformation (yielding) of the composite. It is required that the work performed by the macroscopic stress $\bar{\sigma}_{ij}$ during an incipient plastic deformation process of a representative element be equal to the rate of dissipation $\dot{D}(\dot{\epsilon}_{ij}')$ in the constituents (matrix and fibers) of the composite

$$\bar{\sigma}_{ij}\dot{\epsilon}_{ij} = \frac{1}{V}\int_V \dot{D}(\dot{\epsilon}_{ij}')\,dV \tag{1}$$

where

V = volume of a representative element of the composite,
$\dot{\epsilon}_{ij}$ = macroscopic (average) strain rate, and
$\dot{\epsilon}_{ij}'$ = microstrain rate in the composite constituents.

Radoslaw L. Michalowski, Department of Civil Engineering, The Johns Hopkins University, Baltimore, Md. 21218. Aigen Zhao, Tenax Corp., 4800 East Monument St., Baltimore, Md. 21205.

This technique is similar to using limit analysis, except that here the unknown limit load is the average stress in the composite at failure. A concept similar to this was explored earlier in the context of cementitious composites in works by Hashin (18) and Shu and Rosen (19), and for two-dimensional membranes by McLaughlin and Batterman (20).

A linear macroscopic velocity field v_i is assumed here

$$v_i = a_{ij} x_j \tag{2}$$

where x_j is the Cartesian coordinate and a_{ij} is a matrix of coefficients subject to constraints imposed by the dilatancy of the matrix material. We are considering plane-strain deformation; hence Equation 2 becomes

$$v_1 = -\dot{\epsilon}_{11} x_1 - \dot{\epsilon}_{12} x_2$$
$$v_2 = -\dot{\epsilon}_{21} x_1 - \dot{\epsilon}_{22} x_2 \tag{3}$$

where $\dot{\epsilon}_{ij}$ is the macroscopic strain rate throughout the considered representative element of the composite (compression is taken as positive). Assumption of a linear velocity field has an important consequence: the rate of energy dissipation in yielding fibers depends only on their orientation, not on their particular location in the composite element.

The homogenization technique proposed here is considerably different from that recently suggested in a work by di Prisco and Nova (10), who considered a continuous filament reinforcement. In that approach, the macroscopic stress is arrived at by superposition of the stress in the soil matrix and a fictitious filament structure capable of resisting tension only. It is a statics-based approach similar to one used in the mechanics of mixtures. Although it is a very reasonable method for homogenizing the continuous filament composite, the energy-based technique used in this paper is more convenient when slip of fibers needs to be accounted for.

FAILURE CRITERION FOR FIBER-REINFORCED SAND

Failure of the fiber-reinforced composite occurs when the matrix material reaches the yielding state, which may be associated with slip or tensile collapse of fibers. It is assumed here that the matrix material conforms to the Mohr-Coulomb failure condition and the associative flow rule. The strain rate field must then satisfy the dilatancy relation

$$\frac{\dot{\epsilon}_v}{\dot{\epsilon}_1 - \dot{\epsilon}_3} = -\sin \varphi \tag{4}$$

where $\dot{\epsilon}_v = \dot{\epsilon}_{ii}$ is the volumetric strain rate and φ is the internal friction angle of the matrix, which, for the associative flow rule, also indicates the rate of dilation. Under the assumptions made (Mohr-Coulomb criterion, associativity), the rate of energy dissipation during plastic deformation of a noncohesive matrix is zero (21). Although a purely granular (noncementitious) matrix is considered here, the effort can be easily extended to cementitious (cohesive) matrices.

The use of the associative (normality) rule for soils has been questioned in the past as it predicts unrealistic rates of dilation (Equation 4). The associative flow rule is not unreasonable; however, this issue deserves more space, and it is not addressed here [the reader will find a useful discussion elsewhere (22)].

The amount of fibers is characterized here by their concentration (volume density)

$$\rho = \frac{V_r}{V} \tag{5}$$

where V_r is the volume of the fibers and V is the volume of the entire representative composite element. The yield point of the fibers is σ_0. Fibers contribute to the composite strength only if a tensile force can be mobilized in them. A frictional load transfer mechanism is considered here. An interface shear stress and axial stress in a rigid, perfectly plastic fiber within a uniformly deforming matrix is shown in Figure 1. The maximum tensile stress in fibers (σ_0) can be mobilized only if length l is sufficiently large, otherwise fiber slip occurs. Note that even if the middle part of a fiber is at yield, each of its ends will slip over the distance d (Figure 1).

$$d = \frac{r}{2} \frac{\sigma_0}{\sigma_n \tan \varphi_w} \tag{6}$$

where

σ_n = stress normal to the fiber surface,
φ_w = friction angle of the matrix-fiber interface, and
r = fiber radius.

A uniform distribution of the fiber orientation in three-dimensional space is considered here. The dissipation rate in a single fiber in the deforming matrix depends on the strain in the direction of the fiber. The energy dissipation rate due to fiber tensile collapse was calculated by integrating the dissipation over all the fibers in tension. Contribution of fibers under compression to the strength of the composite is neglected here because of possible buckling and kinking. Because a frictional fiber-matrix interface is assumed in the model, the interface shear stress is dependent on the stress normal to the fibers. Calculations are performed here assuming this stress is equal to the mean of the maximum and minimum stress $p[p = (\bar{\sigma}_1 + \bar{\sigma}_3)/2]$ for all fibers. This assumption leads to a conservative estimate of the energy dissipation rate because the average normal stress on fibers under tension (fibers under compression are excluded in calculations of the dissipation rate) is larger than the mean stress p, and, therefore, the energy dissipation rate due to fiber slip

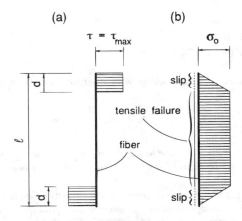

FIGURE 1 Stress distribution on a rigid-plastic fiber in a deforming matrix: (a) interface shear stress; (b) the axial stress.

is underestimated. Consequently, for a uniformly distributed orientation (not uniform orientation) in three-dimensional space, this assumption leads to a conservative estimate of the fiber contribution to strength and is acceptable for such composites. This may not be acceptable for anisotropic patterns of fiber orientation distribution where the mean stress p may be larger than the average normal stress on fibers under tension.

This paper focuses on the concept of macroscopic description itself; the mathematical details of the derivation will not be presented. In short, a plane-strain deformation process of a fictitious composite specimen was considered, and the expression in Equation 1 was used to calculate the macroscopic stress at failure $\bar{\sigma}_{ij}$. It is convenient to represent this failure stress (failure condition) as

$$f = R - F(p) = 0 \quad \text{or} \quad R = F(p) \tag{7}$$

where R is the radius of the limit stress circle, and p is the mean of the extremal principal stresses

$$R = \frac{1}{2}(\bar{\sigma}_1 - \bar{\sigma}_3) = \frac{1}{4}\sqrt{(\bar{\sigma}_x - \bar{\sigma}_y)^2 + 4\tau_{xy}^2}$$
$$p = \frac{1}{2}(\bar{\sigma}_1 + \bar{\sigma}_3) \tag{8}$$

For long fibers where tensile failure can be expected, the failure criterion was found in the form

$$\frac{R}{\rho\sigma_0} = \frac{p}{\rho\sigma_0}\sin\varphi + \frac{1}{3}N\left(1 - \frac{1}{4\rho\eta}\frac{1}{\frac{p}{\rho\sigma_0}\tan\varphi_w}\right) \tag{9}$$

where η is the fiber aspect ratio (r = fiber radius)

$$\eta = \frac{l}{2r} \tag{10}$$

and

$$N = \frac{1}{\pi}\cos\varphi + \left(\frac{1}{2} + \frac{\varphi}{\pi}\right)\sin\varphi \tag{11}$$

When fibers are relatively short ($l \leq 2d$, Figure 1), collapse of the composite is expected to be associated with slip of fibers. This occurs when

$$\eta < \frac{1}{2}\frac{\sigma_0}{p\tan\varphi_w} \tag{12}$$

The failure criterion associated with the slip of fibers takes the form

$$\frac{R}{\rho\sigma_0} = \frac{p}{\rho\sigma_0}\sin\varphi + \frac{1}{3}N\frac{p}{\sigma_0}\eta\tan\varphi_w \tag{13}$$

Note that $\rho\sigma_0$ is used here to normalize the maximum shear stress, and R is independent of the fiber yield stress (σ_0) when pure slip occurs.

The failure criterion is isotropic (uniform distribution of fiber orientation), is independent of the intermediate principal stress (the consequence of the matrix Mohr-Coulomb failure condition), and can be presented conveniently in $\tau_{max} - p$ space ($\tau_{max} = R$). Figure 2 shows the results of calculations for one particular example, where

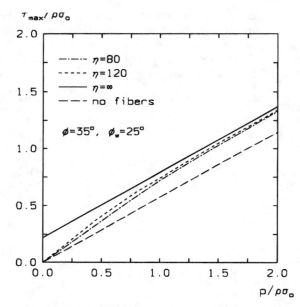

FIGURE 2 Maximum shear stress envelopes for a fiber-reinforced granular material (theoretical result).

constant internal friction angle of the granular matrix $\varphi = 35°$ and angle of fiber-matrix interfacial friction $\varphi_w = 25°$ (this is somewhat higher than what was measured for steel or polyamide). Although the model is sensitive to φ_w, this friction is probably not the single mode of load transfer to the fibers, as will be suggested in the next section. The failure lines are piecewise functions where the range for low p is linear (Equation 13), and it is nonlinear for larger p (Equation 9). Notice that the failure criterion is continuous and smooth (continuous first derivative).

The failure criterion in Equations 9 and 13 is consistent with the constitutive description in the theory of plasticity (irreversible and time-independent behavior of solids) and can be used directly in analytical and numerical techniques for solving boundary value problems in geotechnical engineering. This description is different from the one based on consideration of localized shear during the failure process as proposed elsewhere (17). In the latter, the occurrence of strain localization is predetermined, random distribution of fiber orientation is ignored (all fibers are considered perpendicular to the shear band), and the contribution of fibers to the composite strength is assumed a priori to be a linear function of the confining stress. The increase of the shear strength is then expressed as a function of a distortion angle in the shear zone. The second model also introduces an empirical constant, which further contributes to the fact that the two descriptions cannot be reasonably compared.

PRELIMINARY TEST RESULTS

An experimental program is under way to indicate whether the theoretical description derived is a reasonable characterization of the true failure behavior of fiber-reinforced sand. Comprehensive results from the experimental study are unavailable, but some results are presented in Figure 3. Triaxial tests were performed on specimens of fiber-reinforced sand. It should be pointed out that soils do not conform to the Mohr-Coulomb failure criterion precisely; in particular, yielding of soils is sensitive to the intermediate principal

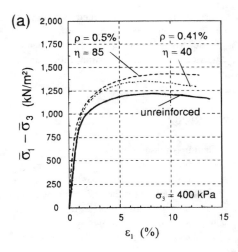

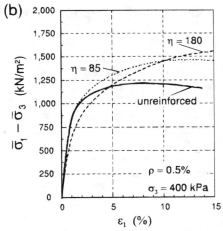

FIGURE 3   Stress-strain curves from drained triaxial tests on fiber-reinforced sand at confining pressure 400 kPa: (a) steel fibers; (b) polyamide fibers.

one to avoid strain localization during tests. Strain localization introduces significant difficulties in interpreting recorded displacements in terms of fiber strains. No strain localization was noticed during the tests performed. The boundary friction effects were minimized by using double layers of silicon grease-lubricated rubber membranes at both ends of the specimens.

Lengths of fibers were close to 2.54 cm (1 in.) in all cases, with the aspect ratio adjusted by varying the diameter of fibers. Each specimen was prepared in five premixed portions to ensure a uniform spatial distribution of fibers. When placing the soil in the mold, the fibers were clearly assuming anisotropic orientation with the horizontal being the preferred direction. Therefore, each part of the specimen was placed over a grid of wires and the grid was pulled through the prepared material, altering the orientation of a portion of the fibers; this led to a nearly isotropic distribution of fiber orientation (the wires in the grid were spread every 3 cm or 1.2 in.). Reinforced and unreinforced specimens were prepared in the same fashion. The nearly uniform distribution of fiber orientation was found by visual inspection, when some specimens were disassembled layer by layer and the fibers were gradually exposed (the matrix was held together by apparent cohesion).

Figure 3 shows the stress-strain behavior of the composite specimens during drained triaxial tests at a confining pressure of 400 kPa. As expected, the limit stress (or the stress at the peak) is larger for a larger aspect ratio, the fiber content being comparable. The stiffness, however, drops considerably when the aspect ratio η of polyamide fibers is increased from 85 to 180, volumetric concentration being 0.5 percent. The internal friction angle of the granular soil at this confining stress was measured to be 37.1 degrees, and the fiber-matrix interface friction angle was 20 and 14 degrees for the steel and polyamide, respectively. The yield point for both materials is roughly 0.7 and 200 MPa; it was not reached in fibers during the tests. Angle φ_w was measured in a test where the monofilament fiber was pulled out from the soil mass placed in a box and also in a direct shear test in which the matrix soil was dragged over a sheet of fiber material. The second test is preferred because the former has inherent uncertainties in interpretation of the stress state in the soil surrounding the fiber.

Figure 3 indicates that, even though the polyamide surface friction angle is less than that for steel, the deviatoric stress at failure for the composite reinforced with polyamide is greater than that reinforced with steel (for a comparable case of $\rho = 0.5$ percent and $\eta = 85$). This demonstrates that the stress induced in the polyamide is larger than that in steel fibers. Polyamide fibers probably did not assume perfectly straight shape after specimen preparation, giving rise to a stress transfer mechanism other than simple friction. Some irreversible flexural deformation was observed on disassembling the specimens, and it was noticed that polyamide fibers also experienced some local damage (but not rupture failure). It was then concluded that it is not the tensile strength of the fibers but the load transfer mechanism that has a detrimental effect on the behavior of the composite.

A comparison of the experimentally derived failure criterion and its theoretical prediction based on the limit analysis homogenization scheme presented in the previous section is shown in Figure 4. This comparison is shown for steel fibers only. It was found in experiments that the polyamide fibers did not retain their straight (linear) shape during tests, but the fibers were assumed to be straight cylinders in the modeling effort. Refinement of the model, including the influence of the fiber flexibility, is yet to be attempted.

stress. The results of plane-strain and axisymmetric tests, therefore, can be expected to be different. It is important for consistency that the internal friction of the soil be obtained from the same type of test that is performed for the fiber-reinforced composite. However, because it was decided that the Mohr-Coulomb condition (independent of the intermediate principal stress) be used for the matrix, the results of the theoretical investigation may be expected not to be sensitive on whether it involves plane-strain or axisymmetrical analysis.

A coarse, poorly graded sand was used, with $d_{50} = 0.89$ mm and uniformity coefficient $C_u = 1.52$; specific gravity of that sand was 2.65, extremal void ratios were 0.56 and 0.89, and the initial void ratio of prepared samples was $e = 0.66$. Steel and polyamide were used as the fiber material. Polyamide is not a material likely to be used as a permanent soil reinforcement (because of deterioration of mechanical properties and moisture sensitivity), but its availability in a variety of sizes, and mechanical behavior common to other synthetics, makes it a convenient material to use in tests.

Triaxial drained tests on specimens of fiber-reinforced soil were performed. The height of specimens was 9.65 cm (3.8 in.) with the diameter-to-height ratio equal to one. This ratio is larger than that for a typical triaxial specimen; it was purposely selected equal to

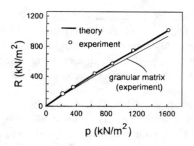

FIGURE 4 Maximum shear stress envelopes for steel fiber-reinforced sand (slip mode of failure).

In the stress range in Figure 4 the fibers did not yield, and the entire energy dissipation occurred from fiber slip in the deforming matrix. The failure condition is nonlinear, however, as a result of accounting for the variation of the matrix internal friction angle with stress (in the range indicated: 43 to 35.9 degrees). Although steel fibers are not a very effective reinforcement, Figure 4 reveals an exceptionally small discrepancy between the experimental and the theoretical results.

EXAMPLE OF APPLICATION

The limit load on a reinforced slope on a firm foundation was calculated using the slip-line method. Two cases are considered: one with fiber reinforcement and a second one with a "traditional" reinforcement (a geogrid, for instance). In both cases the material is homogenized, and the amount of reinforcement is characterized by $\rho\sigma_0$, where ρ is the volumetric fraction of the reinforcement fibers, and σ_0 is the yield stress of the reinforcing material. In the case of unidirectional reinforcement, the macroscopic continuum is, of course, anisotropic (*12*).

The slip-line fields for a slope of inclination angle of 55 degrees and soil internal friction angle of 40 degrees are shown in Figure 5. The aspect ratio of the fibers is 150, interfacial friction angle $\varphi_w = 25$ degrees, and the slope is characterized by dimensionless parameter $\gamma H/\rho\sigma_0 = 0.32$ (γ = unit weight of soil, H = slope height). The average limit load was calculated, and it is given in di-

mensionless fashion: for a fiber-reinforced slope $\overline{q}/\rho\sigma_0 = 2.77$, and for horizontal ("traditional") reinforcement $\overline{q}/\rho\sigma_0 = 10.89$. Such outcome is not surprising because all the reinforcement placed in the horizontal direction is used in tension. Fibers are not all effectively used.

FINAL REMARKS

Potential applications of fiber reinforcement are in infrastructure for transportation, such as subgrades for roads and airfields, embankment slopes, and so forth. Stability analyses of such structures require that the failure criterion for the fiber composite be known. An effort toward describing the stress state at failure of a fibrous granular composite was presented. An energy-based homogenization technique was shown to be a good tool to average the stresses in the composite.

The failure condition was found in the form of two functions, one related to tensile failure of fibers and the other associated with fiber slip. The first represents the shear strength as a nonlinear function of mean stress, whereas the second is linear (unless the variability of the internal friction angle with stress is taken into account). The transition from one failure mode to another is smooth. Neither theoretical nor laboratory test results indicate that the failure condition consists of two piecewise linear segments, as is often suggested.

The failure criterion derived is applicable to fiber-reinforced granular composites where the straight cylinder shape of fibers is preserved during deformation process (such as the steel fiber-reinforced sand tested here). Including the fiber flexibility in the model, with all its consequences to fiber-matrix interaction, is yet to be attempted.

The parameters needed to predict the strength of the fiber-reinforced soil are the soil internal friction angle (φ), volumetric content of fibers (ρ), fiber-soil interface friction angle (φ_w), fiber aspect ratio (η), and the yield point of the fiber material (σ_0). They all have a very clear interpretation.

The failure criterion derived can be used in limit analyses of geotechnical structures, or it can be used in finite element calculations as part of the constitutive model for the composite. Refinement of the theoretical description needs to be directed toward capturing more realistic mechanisms of fiber-matrix interaction.

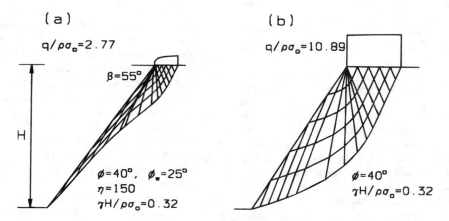

FIGURE 5 Slip-line fields for slope limit load calculations: (a) fiber reinforcement; (b) unidirectional (horizontal) reinforcement.

ACKNOWLEDGMENTS

The work presented in this paper was sponsored by the Air Force Office of Scientific Research, grant No. F49620-93-1-0192, and by the National Science Foundation, under grant No. MSS-9301494. This support is gratefully acknowledged.

REFERENCES

1. Leflaive, E., and Ph. Liausu. The Reinforcement of Soils by Continuous Threads. *Proc., 3rd International Conference on Geotextiles,* Vienna, Austria, Vol. 4, 1986, pp. 1159–1162.
2. Hill, R. A Self-Consistent Mechanics of Composite Materials. *Journal of the Mechanics and Physics of Solids,* Vol. 13, 1965, pp. 213–222.
3. Budiansky, B. On the Elastic Moduli of Some Heterogeneous Materials. *Journal of the Mechanics and Physics of Solids,* Vol. 13, 1965, pp. 223–227.
4. Mori, T., and K. Tanaka. Average Stress in Matrix and Average Elastic Energy of Materials With Misfitting Inclusions. *Acta Metallurgica,* Vol. 21, 1973, pp. 571–574.
5. Dvorak, G. J., M. S. M. Rao, and J. Q. Tarn. Generalized Initial Yield Surfaces for Unidirectional Composites. *Journal of Applied Mechanics,* ASME, Vol. 41, 1974, pp. 249–253.
6. Hashin, Z. Analysis of Composite Materials: A Survey. *Journal of Applied Mechanics,* ASME, Vol. 50, 1983, pp. 481–505.
7. Sawicki, A. Plastic Limit Behavior of Reinforced Earth. *Journal of Geotechnical Engineering,* Vol. 109, No. 7, 1983, pp. 1000–1005.
8. de Buhan, P., R. Mangiavacchi, R. Nova, G. Pellegrini, and J. Salençon. Yield Design of Reinforced Earth Walls by Homogenization Method. *Géotechnique,* Vol. 39, No. 2, 1989, pp. 189–201.
9. de Buhan, P., and L. Siad. Influence of a Soil-Strip Interface Failure Condition on the Yield-Strength of Reinforced Earth. *Computers and Geotechnics,* Vol. 7, Nos. 1 & 2, 1989, pp. 3–18.
10. di Prisco, C., and R. Nova. A Constitutive Model for Soil Reinforced by Continuous Threads. *Geotextiles and Geomembranes,* Vol. 12, 1993, pp. 161–178.
11. Michalowski, R. L., and A. Zhao. Limit Loads on Fiber-Reinforced Earth Structures. *Proc., 13th International Conference of Soil Mechanics and Foundation Engineering,* New Delhi, India, Vol. 2, 1994, pp. 809–812.
12. Michalowski, R. L., and A. Zhao. Failure Criteria for Fibrous Granular Composites. In *Computer Methods and Advances in Geomechanics* (H. J. Siriwardane and M. M. Zaman, eds.), Vol. 2, 1994, pp. 1385–1390.
13. Andersland, O. B., and A. S. Khattak. Shear Strength of Kaolinite/Fiber Soil Mixtures. *Proc., International Conference on Soil Reinforcement: Reinforced Soil and Other Techniques,* Paris, France, Vol. 1, 1979, pp. 11–16.
14. Hoare, D. J. Laboratory Study of Granular Soils Reinforced With Randomly Oriented Discrete Fibers. *Proc., International Conference on Soil Reinforcement: Reinforced Soil and Other Techniques,* Paris, France, Vol. 1, 1979, pp. 47–52.
15. Arenicz, R. M., and R. N. Chowdhury. Laboratory Investigation of Earth Walls Simultaneously Reinforced by Strips and Random Reinforcement. *Geotechnical Testing Journal,* Vol. 11, 1988, pp. 241–247.
16. Gray, D. H., and H. Ohashi. Mechanics of Fiber Reinforcement in Sand. *Journal of Geotechnical Engineering,* Vol. 109, 1983, pp. 335–353.
17. Maher, M. H., and D. H. Gray. Static Response of Sands Reinforced With Randomly Distributed Fibers. *Journal of Geotechnical Engineering,* Vol. 116, 1990, pp. 1661–1677.
18. Hashin, Z. Transverse Strength, In *Evaluation of Filament Reinforced Composites for Aerospace Structural Applications,* NASA CR-207, (N. F. Dow and B. W. Rosen, eds.), 1964, pp. 36–43.
19. Shu, L. S., and B. W. Rosen. Strength of Fiber-Reinforced Composites by Limit Analysis Methods. *Journal of Composite Materials,* Vol. 1, 1967, pp. 366–381.
20. McLaughlin, P. V., and S. C. Batterman. Limit Behavior of Fibrous Materials. *International Journal of Solids and Structures,* Vol. 6, 1970, pp. 1357–1376.
21. Davis, E. H. Theories of Plasticity and the Failure of Soil Masses. In *Soil Mechanics: Selected Topics* (I. K. Lee, ed.), Butterworth, London, England, 1968, pp. 341–380.
22. Drescher, A., and E. Detournay. Limit Load in Translational Failure Mechanisms for Associative and Non-Associative Materials. *Géotechnique,* Vol. 43, No. 3, 1993, pp. 443–456.

Publication of this paper sponsored by Committee on Soil and Rock Properties.

Reinforcement of Fissured Clays by Short Steel Fibers

LUIS E. VALLEJO AND HANKYU YOO

The study involved a laboratory and a theoretical analysis designed to understand whether the use of short fibers increases the shear strength of fissured clays. The laboratory experiments involved the direct shear testing of reinforced and unreinforced clay samples containing a preexisting crack. The theoretical analysis used the principles of linear elastic fracture mechanics theory to determine the direction of crack propagation in the fissured clays. The investigation revealed that, if short steel fibers are added to the fissured clays, their shear strength increased. For the case of a clay sample with a preexisting horizontal crack, the addition of short fibers increased its shear strength by 9 percent. For the case of a clay sample with a preexisting crack inclined at 30 degrees with the horizontal, the shear strength increased by 25 percent with the addition of the fibers. Thus, the use of short fibers appears to be a viable technique for increasing the strength of fissured clays. However, many questions remain about how to effectively reinforce fissured clays in the field, the interaction mechanisms of the fibers with multiple cracks of different lengths and orientations, and the durability of the steel fibers in clay.

Reinforcement of intact clays with short fibers is a viable technique for increasing their shear strength. Studies have demonstrated that the inclusion of short fibers significantly improves the response of intact clays under both static and dynamic loading conditions (1–4). Many clays, however, are not intact but have fissures in their structure (5–9). These fissured clays often form part of earth dams and natural slopes. If these fissured clays are subjected to static or dynamic loads, the fissures propagate and interact in the clays, causing the failure of the slopes and the earth dams (10–12).

This study involved a theoretical and a laboratory investigation designed to answer the question: Can the use of short fibers stop the propagation of fissures in clays subjected to static loads? If the short fibers stop the propagation of the fissures, the strength of the reinforced fissured clays should also increase.

PHASES AND GOAL OF THE STUDY

This study involved two phases. In Phase 1, prismatic samples with induced cracks were tested in a direct shear apparatus to understand the propagation mechanisms of the induced cracks. Linear elastic fracture mechanics theory was used to interpret the types of stresses and the direction of crack propagation in the fissured clay samples. Phase 2 also involved the preparation of prismatic samples of clay with induced cracks However, to this second set of clay samples, short steel fibers were added. The short steel fibers were placed in

University of Pittsburgh, Department of Civil Engineering, 949 Benedum Hall, Pittsburgh, Pa. 15261.

the samples in a direction normal to the propagation direction of the cracks. The direction of crack propagation was obtained in Phase 1 of the testing program.

The findings of this study represent the first step in understanding whether fibers can reinforce fissured clays under controlled conditions in the laboratory. The goal is to implement the laboratory findings in actual field cases.

PHASE 1: FISSURED CLAYS SUBJECTED TO DIRECT SHEAR

Preparation of the Fissured Clay Samples

For the experimental investigation on the effect of fissures in clays with no reinforcement and subjected to direct shear conditions, laboratory-prepared samples of brittle kaolinite with preexisting cracks were used. The kaolinite clay used in the experiments had a liquid limit equal to 58 percent and a plastic limit equal to 28 percent. Dry kaolinite was mixed with distilled water to form a soft soil mass with a water content of about 40 percent. After the mixing, the clay-water mixture was placed in an oedometer 30 cm in diameter and consolidated under a normal pressure of 25.7 kPa for 5 days. After unloading the oedometer, prismatic specimens measuring 7.62 cm long, 7.62 cm wide, and 2.54 cm thick were cut from the clay block.

Immediately after the prismatic samples were cut, when the water content was equal to 30 percent, cracks were artificially made in the samples by inserting and removing thin glass sheets 1 mm thick and 2.5 cm wide in a direction normal to the samples' free face. Two sets of clay samples, each set having a total of three samples, were prepared in the laboratory. One set of clay samples had the induced crack made at 0 degrees with the horizontal (Figure 1a). The second set of clay samples was prepared with the induced crack inclined at 30 degrees with the horizontal (Figure 1b). The samples were allowed to air dry until their average water contents reached a value of about 3 percent. After this was done, the samples were subjected to direct shear loading in the plane stress direct shear apparatus (Figure 2). This apparatus has been described in detail elsewhere (9,13). Under direct shear stress conditions, the cracks in the samples propagated, forming secondary tensile cracks that extended from the tip of the preexisting cracks.

The samples with the horizontal preexisting crack developed secondary cracks that extended in a direction that was inclined at 65 degrees with the plane of the preexisting crack (Figure 3a). The samples with the preexisting crack inclined at 30 degrees with the horizontal developed secondary cracks that followed the direction of the plane of the preexisting crack (Figure 4a). In other words, it

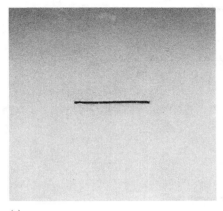

(a)

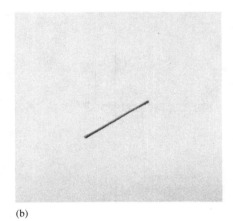

(b)

FIGURE 1 (a) Clay sample with horizontal crack; (b) clay sample with a crack inclined at 30 degrees with horizontal.

extended in a self-similar manner. For the case of the samples with a horizontal crack, crack propagation took place when the normal stress, σ_a, in the direct shear apparatus was equal to 69 kPa, and the shear stress, τ_a, was equal to 380 kPa. For the case of the samples with a crack inclined at 30 degrees with the horizontal, the normal stress, σ_a, in the direct shear apparatus at which the crack propagated was equal to 69 kPa, and the shear stress, τ_a, was equal to 276 kPa.

Theoretical Analysis

To interpret the results obtained from the direct shear results (Figure 3a and 4a), linear elastic fracture mechanics theory was used (*14*). By using this theory, the type of stresses causing the crack propagation in the clay samples as well as the direction of crack propagation can be obtained.

The system of stresses acting on a soil element located on the predetermined location of the failure surface are (*15–17*) those shown in Figure 5, where σ_a and τ_a are the normal and shear stresses exerted on the soil element by the direct shear apparatus, and σ_b is the lateral normal stress to the soil element. Note that in the direct shear apparatus the vertical stress σ_a is equal to the lateral normal stress σ_b (Figure 5).

For the theoretical analysis presented in this study, the soil element in the shear zone as shown in Figure 5 will include an open inclined crack. The soil element will be subjected to the same type of stresses as shown in Figure 5. This system of stresses acts on soil elements located in the shear zone of soil samples subjected to direct shear (*15–17*). The soil element with the crack and the system of stresses acting on it are shown in Figure 6.

In the vicinity of the tip of the preexisting crack (Figure 6), an element in the intact clay that surrounds the crack is subjected to the stresses σ_x, σ_y, and τ_{xy}. According to works by Gdoutos (*18*) and Vallejo (*9*), these stresses can be obtained from

$$\sigma_x = \frac{k_1}{(2r)^{1/2}} \cos \frac{\theta}{2} \left(1 - \sin \frac{\theta}{2} \sin \frac{3\theta}{2} \right)$$
$$- \frac{k_2}{(2r)^{1/2}} \sin \frac{\theta}{2} \left(2 + \cos \frac{\theta}{2} \cos \frac{3\theta}{2} \right) \quad (1)$$

$$\sigma_y = \frac{k_1}{(2r)^{1/2}} \cos \frac{\theta}{2} \left(1 + \sin \frac{\theta}{2} \sin \frac{3\theta}{2} \right)$$
$$+ \frac{k_2}{(2r)^{1/2}} \sin \frac{\theta}{2} \cos \frac{\theta}{2} \cos \frac{3\theta}{2} \quad (2)$$

$$\tau_{xy} = \frac{k_1}{(2r)^{1/2}} \cos \frac{\theta}{2} \sin \frac{\theta}{2} \cos \frac{3\theta}{2}$$
$$+ \frac{k_2}{(2r)^{1/2}} \cos \frac{\theta}{2} \left(1 - \sin \frac{\theta}{2} \sin \frac{3\theta}{2} \right) \quad (3)$$

where r and θ represent the polar coordinates of the point considered (Figure 6), and k_1 and k_2 represent the stress intensity factors (*18*) that can be obtained from

$$k_1 = \sigma_n c^{1/2} \quad (4)$$

and

$$k_2 = \tau_n c^{1/2} \quad (5)$$

In Equations 4 and 5, σ_n is the stress acting normal to the plane of the crack (Figure 6), τ_n is the shear stress acting parallel to the plane of the crack in Figure 6, and c is half the length of the crack. Both of these stresses can be obtained from the following equations.

$$\sigma_n = \frac{\sigma_a + \sigma_b}{2} + \frac{\sigma_a - \sigma_b}{2} \cos (2\alpha) - \tau_a \sin (2\alpha) \quad (6)$$

Since in the direct shear apparatus, $\sigma_a = \sigma_b$ (*15–17*), Equation 6 simplifies to the following relationship:

$$\sigma_n = \sigma_a - \tau_a \sin (2\alpha) \quad (7)$$

The shear stress τ_a can be obtained from

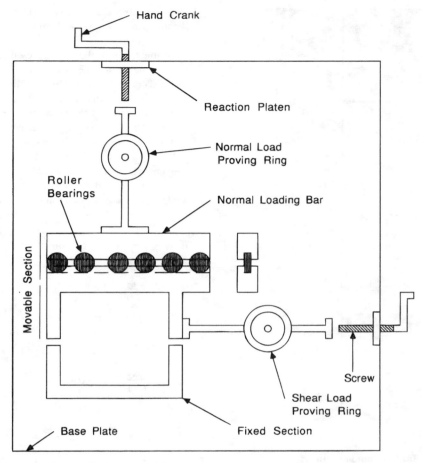

FIGURE 2   Description of the plane stress direct shear apparatus.

$$\tau_n = \frac{\sigma_a - \sigma_b}{2} \sin(2\alpha) + \tau_a \cos(2\alpha) \qquad (8)$$

Since $\sigma_a = \sigma_b$ (Figure 5), Equation 8 becomes

$$\tau_n = \tau_a \cos(2\alpha) \qquad (9)$$

The principal stress σ_1 and σ_3 at points surrounding the crack in Figure 5 can be obtained using Equations 1 through 9 and the following relationship:

$$\sigma_{1,3} = \frac{\sigma_x + \sigma_y}{2} \pm \left[\left(\frac{\sigma_x - \sigma_y}{2} \right)^2 + \tau_{xy}^2 \right]^{1/2} \qquad (10)$$

The direction of the principal stresses can be obtained from the following equations:

$$\psi = \frac{1}{2}\tan^{-1}\left(\frac{2\tau_{xy}}{\sigma_x - \sigma_y} \right) \qquad (11)$$

and

$$\lambda = \psi + \pi/2 \qquad (12)$$

where ψ is the angle of inclination with respect to the X axis (Figure 6) of the principal plane where σ_1 acts; λ represents the inclination with the X axis of the principal plane on which σ_3 acts.

A computer program that uses Equations 9 through 12 was written to calculate and plot the magnitude and direction of the principal stresses around the fissured clay samples shown in Figures 1a, 1b, 3a, and 4a. The principal stresses were calculated using the values of the normal and shear stresses at which the preexisting cracks propagated in the samples.

Analysis of Theoretical Results

Results from the analysis using the computer program are shown in Figures 7 and 8. These figures show the plots of the principal stresses around the horizontal crack and the crack inclined at 30 degrees with the horizontal. The values of σ_a and τ_a that were used to plot Figures 7 and 8 were those at which crack propagation took place when tested in the plane stress direct shear apparatus. For the case of the test that included a sample with a horizontal crack, the values of σ_a and τ_a used were equal to 69 kPa and 380 kPa, respectively. The values of σ_a and τ_a used to plot Figure 8 (the case

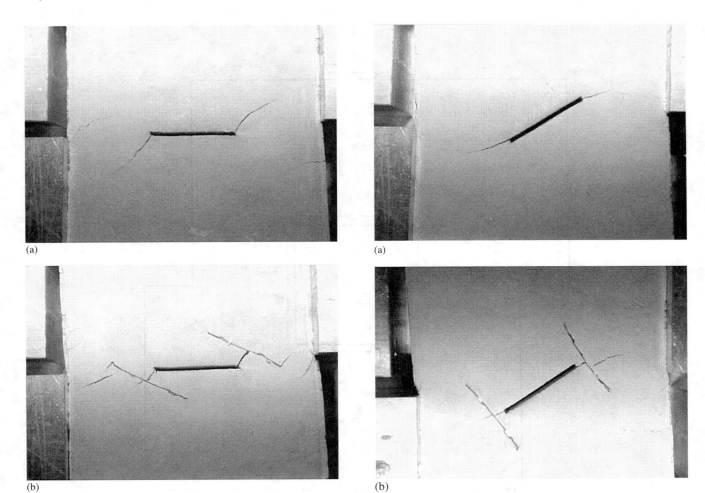

FIGURE 3 (a) Crack propagation in unreinforced clay sample with horizontal crack; (b) crack propagation in reinforced clay with horizontal crack.

FIGURE 4 (a) Crack propagation in unreinforced clay with preexisting crack inclined at 30 degrees with horizontal; (b) crack propagation in reinforced clay sample with preexisting crack inclined at 30 degrees with horizontal.

with the 30-degree crack) were equal to 69 kPa and 276 kPa, respectively. These are the values for the case of clay samples with one crack and no reinforcement.

An analysis of Figures 7 and 8 indicates that zones of tensile stresses formed around the cracks. For the case of the horizontal crack (Figure 7), the zones of tensile stresses formed in the upper-right and lower-left corners of the preexisting crack. These will be the zones in which the soil fails because clays are weak in tension. The laboratory result (Figure 3a) in fact shows that the zones where the tensile stresses developed experienced cracking. The extended cracks will follow a direction that is normal for the major principal tensile stress (Figure 7). This normal direction is equal to 70 degrees (Figure 7) and is very close to the 65 degrees measured in the laboratory experiment (Figure 3a).

For the case of the sample with the preexisting crack inclined at 30 degrees with the horizontal, the computer results are shown in Figure 8. Zones of tensile stresses developed at the two tips of the preexisting crack. The direction of the major principal tensile stress is normal to the plane of the preexisting crack (Figure 8). Therefore,

secondary cracks will extend from the tips of the preexisting crack and will follow a direction parallel to the plane of the preexisting crack. Laboratory results, shown in Figure 4a, confirm the findings of the computer analysis.

PHASE 2: REINFORCED FISSURED CLAYS SUBJECTED TO DIRECT SHEAR

To investigate whether the addition of short fibers is a viable technique to prevent the propagation of cracks in fissured clays, fissured samples similar to the ones shown in Figure 1 (no reinforcement) were prepared in the laboratory. The only difference was that short crimped steel fibers, 2.54 cm long and 0.12 cm wide, were added to the clay (see Figure 9). The short fibers were placed in a direction normal to the expected crack propagation direction (Figures 3a and 4a). The clay samples with no cracks but with steel fibers were subjected in the oedometer to the same level of consolidated pressure (25.7 kPa) for the same duration (5 days) as the samples prepared

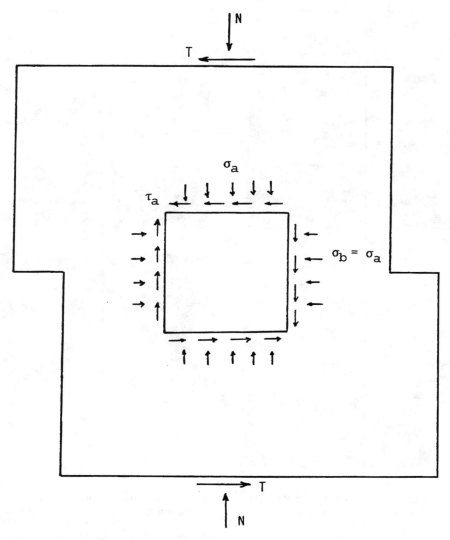

FIGURE 5   Stresses acting on soil element located in shear zone of direct shear apparatus (*15–17*).

for Phase 1 of the testing program. After unloading the oedometer, the cracks were induced in the samples. The steel fibers were not disturbed because their locations were known before the cracks were made in the samples. The samples had a total of eight short steel fibers. The steel fibers were located in two zones, near the tips of the induced cracks, where the secondary cracks developed in the unreinforced fissured clay samples. Each of the two zones had four steel fibers.

The samples with the preexisting fissures were placed in the plane stress direct shear apparatus and were subjected to direct shear. The normal stress, σ_a, in the direct shear apparatus was kept constant and equal to 69 kPa for both samples (one with the horizontal crack, the other with the 30 degree crack). After shearing, secondary cracks developed in both samples. For the case of the sample with the horizontal crack, secondary cracks developed when the shear stress was equal to 414 kPa (Figure 3b). For the case of the reinforced sample with the preexisting crack inclined at 30 degrees with the horizontal, secondary cracks developed when the

shear stress in the direct shear apparatus reached a value equal to 345 kPa.

A summary of the laboratory results on the unreinforced and reinforced samples is shown in Table 1. An analysis of the data indicates that the crimped short steel fibers helped increase the shear resistance of the fissured samples. The addition of the short steel fibers to the sample with a horizontal crack increased its strength by 9 percent. When the fibers are added to the sample with a 30-degree crack, its shear strength increased by 25 percent. Thus, the use of short fibers appears to represent a viable technique for increasing the shear strength of fissured clays.

CONCLUSIONS

The use of fibers to increase the shear strength of *intact* clays has been proved by previous research to be a viable technique. The present study extends the research on reinforced intact clays to include

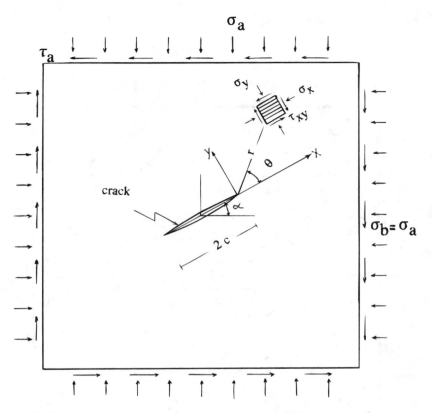

FIGURE 6 System of stresses acting on soil element containing crack and located in shear zone of direct shear apparatus.

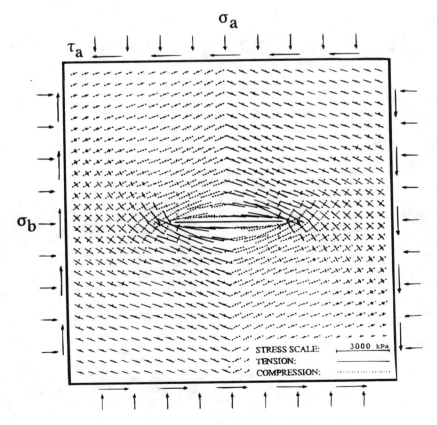

FIGURE 7 Principal stresses around horizontal crack subjected to direct shear stress conditions ($\sigma_a = \sigma_b = 69$ kPa, $\tau_a = 380$ kPa).

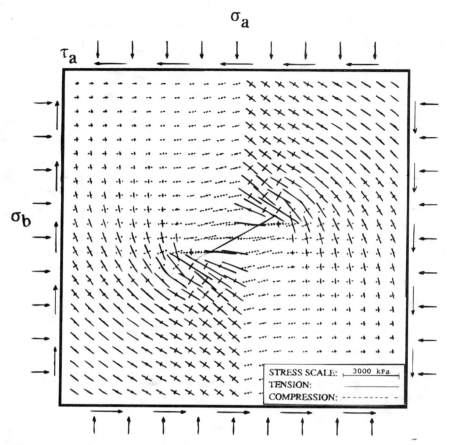

FIGURE 8   Principal stresses around crack inclined at 30 degrees with horizontal and subjected to direct shear stress conditions ($\sigma_a = \sigma_b = 69$ kPa, $\tau_a = 276$ kPa).

fissured clays. By using a laboratory testing program in unreinforced and fissured clay samples, as well as a theoretical analysis that makes use of linear elastic fracture mechanics theory, the following conclusions can be reached.

• The theoretical analysis predicted the type of stresses causing crack propagation and the direction of crack propagation in the clay samples with no reinforcement. The type of stresses that caused the cracks to propagate were tensile in nature, and the direction varied according to the inclination of the cracks in the sample.

• Fibers were added to the samples and were installed in a direction perpendicular to that in which the crack propagated, as determined in the unreinforced fissured clay samples. The addition of short steel fibers to fissured stiff clays increases their resistance to shear stresses. For a sample with a horizontal crack, the addition of the fibers increased its shear strength by about 9 percent. For the sample with a preexisting crack inclined at 30 degrees with the horizontal, the shear strength was increased by 25 percent.

• The short steel fibers appear to be effective in increasing the shear strength of fissured clays if the fibers are placed in the direction normal to the direction of propagation of the preexisting fissures in the clay samples.

• Because the testing program was limited in scope, further research is needed to answer the many important questions that still exist. These questions relate to (a) the repeatability of test results if

different types of clay are used; (b) the effect crack structure, such as crack shape, has on the way cracks propagate in the samples; (c) the influence of the orientation and number of the preexisting cracks on the mechanics of crack propagation and interaction in clay samples; (d) the interaction of fibers and multiple cracks in clay samples; (e)

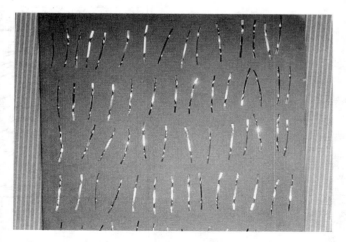

FIGURE 9   Short steel fibers used as reinforcement in fissured clay samples.

TABLE 1   Effect of Reinforcement in Fissured Clays

Sample	Water Content	Normal Stress, σ_a*	Shear Stress, τ_a* (kPa)	
	%	(kPa)	Without Fibers	With Fibers
With Horizontal Crack	3.0	69	380	414
With Crack at 30 degrees with Horizontal	3.0	69	276	345

* σ_a and τ_a are measured by the direct shear apparatus

the methods to introduce the fibers into the clay under field conditions so the fibers are aligned perpendicular to the direction of crack propagation; and (f) the durability of steel fibers in the clay soil.

REFERENCES

1. Andersland, O. B., and A. S. Khattak. Shear Strength of Kaolinite/Fiber Soil Mixture. *Proc., 1st International Conference on Soil Reinforcement*, Paris, France, Vol. 1, 1979, pp. 11–16.
2. Jewell, R. A., and C. J. F. P. Jones. Reinforcement of Clay Soils and Waste Materials. *Proc., 10th International Conference on Soil Mechanics and Foundation Engineering*, Stockholm, Sweden, Vol. 3, 1980, pp. 701–706.
3. Ingold, T. S., and K. S. Miller. Drained Axisymmetric Loading of Reinforced Clay. *Journal of Geotechnical Engineering*, Vol. 109, No. 7, 1983, pp. 883–898.
4. Maher, M. H., and Y. C. Ho. Mechanical Properties of Kaolinite/Fiber Soil Composite. *Journal of Geotechnical Engineering*, Vol. 120, No. 8, 1994, pp. 1381–1393.
5. Terzaghi, K. Stability of Slopes in Natural Clays. *Proc., 1st International Conference on Soil Mechanics and Foundation Engineering*, Cambridge, England, Vol. 1, 1936, pp. 161–165.
6. Duncan, J. M., and P. Dunlop. Slopes in Stiff Fissured Clays and Shales. *Journal of the Soil Mechanics and Foundation Division*, Vol. 95, No. SM2, 1969, pp. 467–491.
7. Lo, K. Y. The Operational Strength of Fissured Clays. *Geotechnique*, Vol. 20, No. 1, 1970, pp. 57–74.
8. Williams, A. A. B., and J. E. Jennings. The In Situ Shear Behavior of Fissured Soils. *Proc., 9th International Conference on Soil Mechanics and Foundation Engineering*, Tokyo, Japan, Vol. 2, 1977, pp. 169–176.
9. Vallejo, L. E. The Influence of Fissures in a Stiff Clay Subjected to Direct Shear. *Geotechnique*, Vol. 37, No. 1, 1987, pp. 69–82.
10. Covarrubias, S. W. *Cracking of Earth and Rockfill Dams*. Harvard Soil Mechanics Series, No. 82. Cambridge, Mass., 1969.
11. Sherard, J. L. Embankment Dam Cracking. In *Embankment Dam Engineering* (R. C. Hirchfeld and S. J. Poulos, eds.), John Wiley and Sons, New York, 1973, pp. 271–353.
12. Vallejo, L. E. Fissure Parameters in Stiff Clays Under Compression. *Journal of Geotechnical Engineering*, Vol. 115, No. 9, 1989, pp. 1303–1317.
13. Vallejo, L. E. A Plane Stress Direct Shear Apparatus for Testing Clays. In *Geotechnical Engineering Congress 1991* (F. G. McLean, D. A. Campbell, and D. W. Harris, eds.), ASCE's Geotechnical Special Publication No. 27, 1991. pp. 851–862.
14. Sih G. C., and H. Liebowitz. Mathematical Theories of Brittle Fracture. In *Fracture* (H. Liebowitz, ed.), Vol. 2, Academic Press, New York, 1968, pp. 67–190.
15. Hill, R. *The Mathematical Theory of Plasticity*. Clarendon Press, Oxford, England, 1950.
16. Hansen, B. Shear Box Tests on Sand. *Proc., 5th International Conference on Soil Mechanics and Foundation Engineering*, Zurich, Switzerland, Vol. 1, 1950, pp. 127–131.
17. Morgenstern, N. R., and J. S. Tchalenko. Microscopic Structures in Kaolin Subjected to Direct Shear. *Geotechnique*, Vol. 17, No. 2, 1967, pp. 309–328.
18. Gdoutos, E. E. *Problems of Mixed Mode Crack Propagation*. Nijhoff Publishers, The Hague, The Netherlands, 1984.

Publication of this paper sponsored by Committee on Soil and Rock Properties.